T0133032

# SENTIMENTAL SAVANTS

# SENTIMENTAL SAVANTS

Philosophical Families in Enlightenment France

MEGHAN K. ROBERTS

THE UNIVERSITY OF CHICAGO PRESS

CHICAGO AND LONDON

MEGHAN K. ROBERTS is assistant professor of history at Bowdoin College.

The University of Chicago Press, Chicago 60637
The University of Chicago Press, Ltd., London
© 2016 by The University of Chicago
All rights reserved. Published 2016.
Printed in the United States of America

25 24 23 22 21 20 19 18 17 16     1 2 3 4 5

ISBN-13: 978-0-226-38411-5 (cloth)
ISBN-13: 978-0-226-38425-2 (e-book)
DOI: 10.7208/chicago/9780226384252.001.0001

Library of Congress Cataloging-in-Publication Data

Names: Roberts, Meghan K., author.
Title: Sentimental savants : philosophical families in Enlightenment France /
    Meghan K. Roberts.
Description: Chicago ; London : The University of Chicago Press, 2016. | Includes
    bibliographical references and index.
Identifiers: LCCN 2015042753 | ISBN 9780226384115 (cloth : alk. paper) |
    ISBN 9780226384252 (e-book)
Subjects: LCSH: Families—France—Philosophy. | Philosophers—Family
    relationships—France. | Enlightenment—France.
Classification: LCC HQ623.R63 2016 | DDC 306.85—dc23 LC record available at
    http://lccn.loc.gov/2015042753

♾ This paper meets the requirements of ANSI/NISO Z39.48-1992 (Permanence of
Paper).

# CONTENTS

What did it mean to be a philosopher in the Age of Enlightenment? Was the ideal man of letters, as tradition held, a celibate bachelor married to Philosophy? Or was he a man of the world, married, possibly with children? These questions provoked much hand-wringing during the eighteenth century. Men of letters aspired to be sociable and useful but also wanted to safeguard their independence. Would polite socializing help them communicate with each other and accrue new knowledge? What was the value of undisturbed solitude? If marriage was healthier and more socially useful than celibacy, should anyone sign up for a life of sexless bachelorhood? These questions were, in many ways, old ones: similar debates had flared up during the Renaissance and Reformation. But the savants of the French Enlightenment attacked these questions with zeal, and a significant number decided they could best live their intellectual and social ideals by marrying and fathering children. Many founded families, a move that transformed their work in fundamental ways.

Families, both real and metaphoric, occupied considerable real estate in the cultural landscape of eighteenth-century France. Family metaphors constituted the bedrock of political authority, with the king as father of his people. Novels, many of them coated in a thick gloss of family feeling, became wildly popular. Plays dramatized family stories with bold gestures and copious weeping. Treatises proposed new ways to nurture family emotions, seeing the love between husband and wife, mother and child, father and family as key to social reform. A thriving public sphere remained interwoven with more intimate categories. Stories about individual families functioned as metaphors for social corruption and regeneration, while revolutionaries imagined private life as a font of public virtue and patriotism. Families were everywhere in the print culture of the period, and the philosophes of the Enlightenment peddled family feeling with abandon. They authored sentimental texts and wept over novels. Family stories plainly appealed to them.[1]

And yet, for all that historians know about domestic life, for all that scholars have contemplated private life and its public import, few have studied the stories that men of letters told about their own family lives. Why not? The question of married men of letters has attracted little attention from historians in part because they have assumed that conspicuous cases—Rousseau's abandoned children, Voltaire's affairs—speak for all savants. Such examples have encouraged historians to see family life as removed from the work of philosophy. In *The Literary Underground of the Old Regime*, for example, Robert Darnton suggests that the true philosophe could only be an unmarried man. A philosophe's marriage signified his domestication, his capitulation to tradition and conformity. Nor is Darnton the only historian to assume that thinkers cared little for marriage. Only a few scholars, such as Anne Vila, have explored questions related to married savants.[2] Biographies abound, and these do examine a thinker's family life in rich detail.[3] Given their focus on one individual, however, biographies are not the best medium for discovering wider trends in Enlightenment culture. When viewed one at a time, married philosophers seem exceptional rather than a part of a larger cultural pattern. And so for most scholars, families have rarely factored into the intellectual histories of the eighteenth century. The myth that family life had little to do with philosophy has persisted.

*Sentimental Savants* solves this problem of scale by drawing on a number of case studies: the Du Châtelet, Diderot, Helvétius, d'Epinay, Condorcet, Necker, Lavoisier, and Lalande families. I am hardly the first scholar to study the likes of these, but my approach connects these famous individuals to their culture in new ways. By examining savants immersed in a diverse array of fields, from chemistry to pedagogy, agronomy to anatomy, I show that shared cultural practices united them. Rather than separating natural philosophers from novelists, women from men, nobles from bourgeois, I look instead at their common traditions and ideals. This is not to deny that real differences existed between my case studies. The savants studied here had varying levels of wealth and social privilege, and I take care to note the ways in which those circumstances shaped their individual experiences.[4] But despite those differences, many philosophes gravitated toward similar forms of self-representation and intimate empiricism in their correspondence and published work. To further demonstrate the wide appeal of these ideas and practices, I glean evidence from collective sources such as eulogies and biographical compendiums.

Using this mix of sources, I argue that a new intellectual ideal emerged in the eighteenth century. Bachelorhood (if not necessarily celibacy) continued to appeal, but family life also beckoned. How lovely it would be, thought some philosophers, to have a devoted wife and affectionate children. Seduced by visions of domestic bliss and familial collaboration, many men of letters turned away from an ascetic life of bachelorhood and dreamed of other possibilities. A new model captivated them: a brilliant man whose loving wife and devoted children enhanced, rather than distracted from, his intellectual life. Eighteenth-century thinkers had come a long way from the curt dismissal of domesticity issued by the scholastic philosopher Peter Abelard (d. 1142), who had asked: "What could there be in common between scholars and wetnurses, writing desks and cradles, books, writing tablets, and distaffs, pens and spindles?"[5] Rather than defining the life of the mind as a total retreat from domestic life, married Enlightenment philosophers imagined social bonds and family love as nurturing and supporting their work. Portraying themselves as loving husbands and fathers—the very picture of idealized sentimental masculinity—helped thinkers represent themselves as engaged members of society who provided a virtuous example for the public to follow.

These sentimental savants, as I call them, were interested in more than emotional solace. An exciting new array of empirical possibilities called to them. Married philosophes often had children, and those children offered a tantalizing opportunity for savants to practice an intimate form of empiricism. Education, so vital to Enlightenment thought, need not be an abstract exercise; they could try out the latest theories. Likewise, philosophes who advocated smallpox inoculation did not need to confine their enthusiasm to the pages of their pamphlets. They inoculated their children, generating proof for the public that the technique was indeed safe and effective. Philosophes seasoned their publications with references to these private practices, bolstering their intellectual authority and anchoring their ideas in compelling narratives. They invited other parents to follow their example, to see savants as moral exemplars as well as extraordinary intellects. If more and more families made themselves like philosophes' families, they suggested, social reform would take deeper root.

Intellectuals had married and had children before the eighteenth century, but during the age of sentiment, domestic narratives were infused with new importance. Loving family ties were the foundation of natural virtue, sociability, patriotism, and social utility. By emphasizing how devoted they were to their families, men of letters tapped into fashionable ideals of sentimental family life and portrayed themselves as exceptionally good men deserving of public trust and emulation. Did every one of them have the idyllic family life they claimed to have? Probably not, although it is difficult to tell from existing sources. But even if social reality fell short of the sentimental ideal, these savants still participated in a new intellectual practice that encouraged them to wear their hearts on their sleeves and turn up the dial on sentimental affect. That they loved their wives and cared about their children became a part of their public personae. It signaled to the public that these men were trustworthy, sensitive, and useful. Marriage was not something to hide: it was an asset to flaunt. In a world where philosophers competed fiercely for audiences, stipends, and general renown, married men of letters used their domestic experiences to give themselves an edge, to fashion themselves as experts on a host of topics.

Indeed, philosophes had great ambitions and hoped to remake their society. Fairly or not, many Enlightenment thinkers lampooned previous generations of scholars—not to mention their present-day enemies—for being aloof, useless, and focused on silly quibbles.[6] By contrast, they represented themselves as fully engaged with society. That they believed they could and should transform their world is significant in and of itself. Older cosmologies had represented society as divinely ordained or historically determined; past philosophers focused their reforms on reversing decline and returning society to a more perfect order created by God and their ancestors. Eighteenth-century philosophers, however, saw society as man-made, with many believing that human beings themselves were at least partially malleable.[7] Many individuals felt confident in their ability to exact social and political change.

This new understanding of society inflected much Enlightenment thought and encouraged men of letters to use their families as propaganda in their efforts for social reform. The potential implications of domestic experiments were thus far reaching. Because family metaphors informed the way people understood society, reforming the domestic sphere could have implications outside intimate spaces. Savants saw family life as a laboratory in which they could test new ideas *and* as a model of society writ small. Philosophes were not simply proposing new ways of organizing families; they were proposing new ways of organizing society.

Marriage and family life were far from peripheral to the making of knowledge in Enlightenment France. The family lives of eighteenth-century savants belong at the center of the Enlightenment, not in the realm of biographical ephemera, because marriage and family life became important tools for the practice and promotion of ideas. Instead of dismissing wives as silly, stupid, or shrewish, savants celebrated their marriages as loving and productive. In lieu of conceptualizing their studies as something best done alone, savants trained their spouses and offspring to act as their assistants. And instead of treating their intimate lives and intellectual work as wholly distinct, these thinkers transformed their families into the object of their investigations. They turned their children into test subjects for their social and medical theories. For a host of reasons, then, the family was a crucial institution of the Enlightenment. The conspicuous display of family love became a key method by which savants positioned themselves as worthy and virtuous leaders of the public, and the use of the family home as an experimental space provided an important new venue for social and scientific experiments.

Family life was about more than rhetoric, of course, and I also aim to reconstruct the experience of domestic life: who worked with whom, who did what with their children. Or, at least, I reconstruct these domestic scenes as well as the passage of 250 years permits. Like many historians of family life, I often found myself exasperated while conducting my research, wishing that more of my subjects had paused to sit down and record their daily lives in minute detail, with attention to how they felt and why (and, if you please, on thick paper and in beautiful handwriting). But most did not do this on a regular basis, and I could steal only the occasional glance into a drawing room or laboratory. Put together, however, these scenes help construct a more comprehensive look into intellectual family life.

At heart, however, this remains a book about stories: what philosophes told their public about themselves and their families. Savants used personal anecdotes to cultivate favorable reputations and to assure their audience that they had personal experience with matters like education and inoculation. These stories were central to the philosophes' attempt to position themselves as moral leaders and figures worthy of emulation. Eager to attract public attention and have a real-life impact, many thinkers longed to break out of their studies and influence their readers' lives. Stories about families helped them do that. This is not to say that the social reality of family life did not matter—it most certainly did. Whether or not savants truly lived up to the sentimental ideals they espoused, they nevertheless had actual wives and children and so could write about families in concrete fashion. They could unite the language of feeling with a rhetoric of empiricism, a brew that made for a potent blend. Their rhetoric drew strength from the fact that they seemed to be practicing what they preached. The pages that follow are thus not strictly intellectual history, or social history, or cultural history, but a combination thereof. This is a history of eighteenth-century ideas as they intersected with cultural, social, and epistemological practices.

At the same time, basing authority on personal experience and natural virtue helped equalize expertise in eighteenth-century France. Male savants were not alone in their ability to observe their children or write moving stories; women of letters could engage in similar work, and I am interested in their stories as well. Indeed, claims of personal experience were so accessible that they fragmented expertise, especially on controversial topics like inoculation. The story of family life and intellectual authority is not one of straightforward progression or stability. As savants traversed the shifting topography of intellectual ideals, they had to continuously reinforce their personal authority, to tell new stories about family life that would give them the upper hand over their detractors.

Given the previous pages' emphasis on family love as intellectually valuable, it will come as no surprise that I find much fault with labeling the Enlightenment the "Age of Reason." In this, I have good company. Historians and literary scholars have done much to dismantle the narrative of the Enlightenment as an era of rational progress, led by philosophes devoted to reason and reason alone. Scholars no longer see emotions as cordoned off from the Enlightenment. Instead, sentiment was central to the intellectual culture of the eighteenth century, even in scientific fields.[8] The Age of Reason, scholars now argue, was also an age of sensibility: reason and emotion were both critical to the project of enlightenment. Men and women admired reason and emotion in equal turn, and they aspired to live their lives as rational but sensitive individuals; sensibility was a way of life.[9] Diderot and D'Alembert, editors of the grand *Encyclopédie* that sought to model human reason and catalogue all knowledge, both wept over sentimental novels. The title "Age of Reason" only captures part of the Enlightenment.

The term is further problematic in that it suggests a single unified, pan-European Enlightenment. An older historiography of the Enlightenment, exemplified by the works of Ernst Cassirer and Peter Gay, defined the movement as a discrete set of ideas and debates. To be an Enlightenment philosophe entailed accepting key dogma. A few historians, such as Jonathan Israel, continue to define the Enlightenment as a kind of catechism, focusing on religious skepticism and radical political philosophies.[10] But most scholars now argue that the Enlightenment lacked stability and only acquired coherence retrospectively, when conservative and radical politicians looked back on eighteenth-century thought to explain the causes of the French Revolution.[11] Enlightenment thinkers advanced many different, sometimes contradictory, ideas. Nor were philosophers united by shared convictions. Some were devoutly religious, some were not, and you can spot similar disjunctures in any area of thought.

In part, the onetime coherence of the Enlightenment has fallen apart because the number of thinkers and modes of intellectual inquiry now associated with it has grown tremendously. Historians once studied a small group of thinkers as *the* Enlightenment, but now a huge number of women, artisans, statesmen, and clerics crowd the stage.[12] Due to this capacious understanding of who counts as an Enlightenment thinker, the movement now looks like a much wider and, correspondingly, looser phenomenon than historians once thought.

When historians pull their focus outward from the most famous philosophes and view the Enlightenment from this wide angle, it becomes impossible to define the movement as a cohesive set of beliefs. Instead, many scholars focus on the development of new cultural and intellectual practices as characteristic features of the Enlightenment. Thinkers espoused many different ideas, but they broadly shared a commitment to public debate and sociability. These practices took many forms: gathering in salons, submitting essays to academic prize contests and newspapers, joining Freemason lodges, planning daily meals, and so forth.[13] Enlightenment actors aspired to find new ways to engage with and reform their society, which spawned new social and cultural practices. Thinkers may not have agreed on their ideas, but they shared new modes of engagement with and for society. Indeed, "the social" is a particularly fruitful concept for the study of eighteenth-century intellectual culture. This angle allows historians to shift from defining "the" Enlightenment to considering how individuals imagined they were *living* the Enlightenment.[14] By focusing on shared practices of self-fashioning and domestic rhetoric, *Sentimental Savants* takes a similar approach.

Focusing on cultural practices has already reaped bountiful returns for studies of the French Revolution. Revolutionaries did not limit themselves to reforming their government but instead conceived of the Revolution as an all-encompassing affair, with total social transformation being their ultimate goal. They tackled subjects as seemingly disparate as fashion and families, religion and political representation, social hierarchy and education.[15] They threw out old ideas of society and embraced the new with much enthusiasm. No aspect of social life stood outside their purview. During the Revolution's most radical phase, political figures even rejected the Gregorian calendar in favor of a secular calendar composed of ten-day weeks. Revolutionaries went beyond Enlightenment conceptions of reform: this was society building on steroids. But it is still useful for scholars to consider the roots of these revolutionary social visions in eighteenth-century culture and to consider how Enlightenment families took their own steps to revamp society.

By focusing on the intersection between the language of feeling and intellectual practice, I also contribute to the burgeoning field of the history of emotions, which traces how the expression and experience of emotions has changed over time. William Reddy has pioneered new understandings of emotions in eighteenth-century France, focusing his efforts on uncovering "emotional regimes" (systems of emotional norms that shore up social hierarchies).[16] Reddy argues that, in the middle of the eighteenth century, elites embraced sentimental outbursts of emotion as they sought to break free from the strictures of court life. My work likewise explores the shift toward sentimentalism as well as its consequences, but in a different setting.

My arguments have also been much informed by the history of science, especially those historians who work on the social construction of intellectual authority. Scientific "truths" did not just spring forth from the heads of lone geniuses nor did those geniuses enjoy automatic support. Rather, their authority—their expertise—had to be socially constructed. They had to represent themselves as using established methods, aligned with accepted theories, affiliated with prestigious institutions, or associated with social elites. To take an English example, when Robert Boyle and Thomas Hobbes went head to head over the moral and scientific value of constructing an air pump, it helped Boyle's case that he was a gentleman associated with the Royal Academy and that Hobbes was not.[17] Claims to scientific knowledge were not made in a vacuum, so to speak.

By the eighteenth century, emotion had become one of the key characteristics of scientific authority. French savants stressed emotion as the wellspring of knowledge, making sentiment a cornerstone of their credibility. These were no dispassionate observers. Instead, thinkers presented themselves as steeped in emotion, social connections, and empirical knowledge.[18] The modern idea of science as a purely rational, detached exercise was nowhere to be found in early empiricism. Studying philosophes as family men reveals new ways in which eighteenth-century savants constructed their personal authority and developed empirical knowledge. Virtuous social attachments enhanced their credibility and enabled empirical investigations, in an appropriately sentimental fashion.

Writing the history of these families shows that many actors contributed to scientific work, itself a major theme in the history of science. Studying "science" in the early modern period—a time when few people studied nature in a professional capacity and when the term "scientist" had yet to be coined—reveals a veritable chorus of collaborators: artisans, barber surgeons, and jewelers, just to name a few.[19] Women's historians have been especially successful in revealing the hidden labor of individuals inside and outside scientific institutions, work that had been erased from traditional accounts of the early modern period.[20] Women participated in intellectual life in many ways: they were authors trained by their doting fathers, well-heeled gentlewomen who studied botany, patrons of learned individuals, and astronomers who collaborated with their husbands, fathers, and brothers. *Sentimental Savants* is my attempt to dispel the old notion of scientists and philosophers as solitary geniuses whose families had nothing to do with their intellectual success. Wives, daughters, sons, brothers, and sisters all worked together to develop new knowledge. I build on earlier scholarship by exploring new case studies, such as the family of the astronomer Jérôme Lalande.

But I am also interested in pushing the scholarly conversation about the domestic context of science in new directions by focusing on reputation. The family functioned as a self-fashioning tool for men and women. Female associates cropped up in correspondence, were acknowledged in the prefaces to published works, and were featured in works of art as loving and learned collaborators. Men, too, draped themselves in a mantle of family love as a part of their public self-fashioning. *Sentimental Savants* not only reveals the work done by family members but also explores new methods of asserting intellectual authority, new public roles for philosophers, and the use of the domestic sphere as a testing ground for larger social reforms. In studying the family as a creative space for making knowledge and building reputations, I am not claiming that family life was entirely beneficial for all family members all the time. The relative benefits of intellectual households were especially complicated for women. Family life opened certain doors for female savants and afforded them recognition for their work. But it also imposed limitations. In most cases, women worked in relation to a male family member and did not pursue autonomous careers. They did not determine their own schedules or set their own agendas. Accordingly, some scholars suggest that the limitations of family workshops for women may have slowed the rise of women working as scientists in their own right. Intellectual households did indeed have their constraints and they generally failed to help women develop independent careers, but they also introduced more women to scientific study and made their involvement visible to savants in the Republic of Letters. By discussing these opportunities, I am not dismissing feminist critiques of scholarly households. Family life was not always good for women, but it was not wholly bad either.

In my focus on family as a testing ground and a site for self-fashioning, I join those scholars who show that many intellectuals developed ideas and staged their personae in the domestic sphere. Sarah Ross's study of Renaissance women, Paul White's work on Darwin, Deborah Coen's study of the Exner family, and John Randolph's research on the Bakunin family all explore different contexts, and yet together reveal that families contributed to the production of knowledge in varied ways.[21] These studies show we cannot assume that neat barriers divided public from private. Instead, the public was invited to peek into the windows of the family home, to admire the ties that bound family members, and to emulate the conduct and methods of learned families. Enlightenment France has not yet been a big part of the history of intellectual households and scientific reputations, but the eighteenth century—an age of sentiment, and a period when public and private intersected in powerful ways—was a key phase of this history.[22]

Some of these works—Coen's, White's, and Randolph's—focus on a single family. I have chosen to follow a different model, however, more akin to Sarah Ross's "experiment in collective biography," by studying a range of eighteenth-century thinkers from a variety of social milieux. Although the *gens de lettres* studied here hailed from different backgrounds, they were not unconnected. Many of them knew each other: some belonged to the same academy, others socialized together at salons, and many had mutual acquaintances. They belonged to the same "emotional community," a group of people bound together by shared texts, traditions, or living spaces that have a common "system of feeling."[23] They read many of the same works, they adhered to many of the same ideals, and they practiced similar emotional styles. They shared a common intellectual culture and engaged in similar forms of intellectual exchange, even if the content of their ideas varied.

Yet the savants discussed in the following chapters represent a wide range of intellectual and cultural activities, and the study of so mixed a group poses a few problems. The savants studied here are sufficiently diverse to resist many of the traditional titles applied to Enlightenment thinkers. The term "philosophe" does not fit all my case studies, as it referred to a particular set of thinkers associated with Voltaire or the *En-*

*cyclopédie*. I more often use inclusive titles such as *gens de lettres* and "savants" while restricting the term "philosophe" for those associated with that *parti*, such as Diderot and D'Alembert. *Gens de lettres* (or, in its masculine version, "men of letters") and "savants" refer more generally to those engaged in intellectual work, and are more universally applicable to the subjects studied here. All these terms were contested and ever-shifting in their meaning. Rather than using labels to draw boundaries between thinkers, I employ "savants" and *gens de lettres* to suggest people who self-identified as engaging in "enlightenment," whatever that meant for them.

The families considered in this book make for a varied group, but they are unified in one respect: they are all French. In studying the Enlightenment in France, however, I am not claiming that the Enlightenment was essentially French. Work on the Enlightenment in European, American, and global contexts makes clear that it was not.[24] Nor am I arguing that well-known philosophes like Diderot and savants like Lavoisier are more worthy of study than lesser-known individuals. Mine is just one context of many. I hope that scholars will pick up this project in other milieux to develop a fuller understanding of intellectual families.

Also worthy of future research are nonbiological families. An enormous variety of intellectual households proliferated in eighteenth-century France. I have chosen to focus on biological families united by marriage because that was the household that prompted the most discussion: the most anxiety among those who worried about philosophers compromising their impartiality, the most praise among those who saw marriage and fatherhood as a man's natural obligations. Few philosophers warned against the dangers of a teacher treating his students like his sons; they worried far more about the financial burden and everyday distractions of small children. Likewise, same-sex households—with two male philosophers sharing one roof—prompted little anxiety. Male friendship of this sort was highly praised. Other philosophers, like D'Alembert, lived with women who were not their wives. Focusing on debates about marriage and biological fatherhood is not an attempt to privilege heteronormative families over other household arrangements. There is still much to study with regard to savants and their households; I have only covered some of the terrain here.

The book begins with debates about philosophers marrying and having children: Would family life make them better or worse philosophers? More useful to society or useless as thinkers? I then explore how family life and sentimental self-fashioning shaped the making of Enlightenment knowledge: how wives and children collaborated with savants to observe, calculate, and promote new works; how the family became a laboratory for testing and advertising ideas about education and inoculation. These chapters look at flash points in the history of intellectual family life, with each spanning the eighteenth century; many events discussed in separate chapters unfolded around the same time. Together, these chapters reveal family life as the stage on which savants played their parts.

Philosophes' families were *sui generis* in many ways, but in revealing new links between public and private, emotional and intellectual life, they have a broad historical significance. They remind us that the Enlightenment was a complex blend of ideas for reform and renewal. Eighteenth-century individuals did not want sterile debates. They wanted to imagine new ways to live, new ways to build society. This desire to live the Enlightenment did not stop at the family doorstep. Rather, the family home was fertile ground for developing new ideas, new social theories, and new cultural practices. Families—husbands, wives, daughters, and sons—were key players during the Enlightenment. Men and women aspired to represent their families as loving, learned, and respectable. By drawing attention to the ordinary virtues of private life, they opened up debates about the relevance of personal virtue to public authority. Those debates reverberate still today.

# CHAPTER ONE

# Men of Letters, Men of Feeling

In 1751, Jean-Jacques Rousseau received a letter from a horrified reader. Suzanne Dupin de Francueil had heard a rumor that Rousseau had abandoned five children at foundling hospitals: Could this be true? Defiant, Rousseau acknowledged that he was, in fact, guilty as charged but insisted that he had had no other choice. It would have been impossible for him to be father and philosopher at the same time. Raising a family required a constant flow of cash, and that was not easy to come by as a writer. If he had kept his children, he would have been reduced "to intrigue, to games, to craving some vile employment, to bettering himself by the ordinary means." Repulsed by this life of craven dependence, Rousseau swore that "it would be better for my children to be orphans than to have such a rascal for a father."[1] He saw family life and philosophy in diametric opposition, and he could not have pursued both at the same time.

The story of Rousseau's abandoned children was and is notorious: How could the author of *Emile*, an educational treatise that demanded parents devote themselves unswervingly to their children, and *La Nouvelle Héloïse*, a novel that made legions of readers weep over the joys and sorrows of love, have done such a thing? On its own, Rousseau's example suggests that Enlightenment philosophers avoided family life at all costs and devoted themselves to lives of stoic contemplation rather than sentimental connection. Historians of eighteenth-century France have long known, however, that the life of Jean-Jacques Rousseau—an acerbic and eccentric individual—was far from normal. Indeed, in surveying a wider swath of French thinkers, I have discovered men of letters quite unlike Rousseau. These men did not shun family ties but rather made them part of their public personae. Some very well-known philosophes—including Claude-

Adrien Helvétius, the Marquis de Condorcet, and Antoine Lavoisier—
count among these numbers.

Rather than representing themselves as independent and unattached,
these men of letters embraced more sentimental modes of self-fashioning.
They depicted themselves as men of letters *and* men of feeling. They
crafted a new intellectual ideal, that of a married couple who loved each
other deeply and collaborated productively. The rise of this ideal marked
the creation of a new type of public man, a figure I am calling the "sentimental savant": a philosopher who immersed himself in family feeling but
retained his intellectual edge. Inspired by the treacly prose of sentimental
novels, many men of letters described family life in over-the-top language
and presented themselves as loving husbands and fathers. Pierre Choderlos de Laclos, author of *Dangerous Liaisons*, rhapsodized that, "whatever
might become of me, I console myself with the idea that you [my wife]
will be my posterity, and that my memory will find asylum in your heart.
The pure and sensitive heart of a good wife and a good mother is a pan-
theon worth as much as any other."[2] Even bachelor savants emphasized
that they still had families and that they cared for them deeply.[3] Drawing
on the language of sentimental domesticity became one way to prove virtue, sensitivity, and trustworthiness. Enlightenment philosophes thought
of themselves as public figures, and representing themselves as loving husbands and fathers allowed them to claim to lead by example.

Thinkers had married and had children before the Enlightenment,
and so the major element of change over time studied here is not a demo-
graphic shift but rather a shift in cultural ideals and representations. With
philosophes increasingly anxious to present themselves as useful and patriotic members of society, as moral leaders who would reform the public
by providing a model of behavior, they deployed the language of feeling as
a way to stress their virtue and sociability. Highlighting their affectionate
domestic lives helped accomplish this goal. The ideal of the happily married man of letters, the philosophe as devoted father, became a new way
(but not the only way) for savants to represent themselves as engaged and
admirable citizens.

This new image of intellectual life coexisted alongside other models.
Many Enlightenment philosophers—including clerics who had no choice
in the matter—remained unmarried, and without controversy. Like their
married counterparts, bachelor men of letters could claim to be sociable,
virtuous, and public-minded. The philosophe *par excellence*, Voltaire,
did not consider his unmarried status to be a black mark on his charac-

ter. Indeed, even writers who lavished praise on loving marriages could marvel in turn at a celibate savant devoted to his work. Enlightenment thinkers thus had choices to make, even as many embraced sentimental domesticity.

They did so from across the social spectrum, although most of the thinkers discussed here came from the middling and aristocratic levels of society. Their comfortable socioeconomic statuses undoubtedly made the decision to wed easier for them, and their exposure to sentimental texts clearly shaped their self-representations. Nevertheless, I have not found material factors in and of themselves sufficient to explain changing ideals of intellectual marriage.[4] While social status clearly mattered in both the practice and representation of domestic life, it is difficult to pinpoint a clear social origin for the ideal of companionate intellectual marriage. It would be wrong to say that the ideal of the sentimental savant was fundamentally bourgeois.[5] The language of feeling was picked up in a scattershot way by savants from a range of social backgrounds and intellectual interests. To deal with this heterogeneity, I have chosen to restrict my observations about class to individual examples rather than generalizing from them.

Savants may have talked the talk of sentimental family life, but they did not always walk the walk. Clearly not—Rousseau was, after all, the master of the sentimental novel, a trusted source of child-rearing advice, and seemingly everyone's *ami*, yet that did not stop him from abandoning his own children. The men and women discussed in this chapter did not reach that level of hypocrisy, but contradictions aplenty can be found in their lives and letters. Helvétius wrote honeyed letters to his wife, and yet he frequented prostitutes. Marie-Anne Lavoisier, whose public image was that of devoted wife and attentive assistant, had a long-term affair with her husband's friend Samuel Pierre Dupont de Nemours. Even beyond these striking examples, families experienced tensions and betrayals that did not jibe with the harmonious and affectionate image they put forward. Some savants clung to aristocratic licentiousness even as they embraced new forms of sentimental domesticity; others displayed ordinary human changeability. These individuals may have been sincere when they expressed marital affection and familial devotion, but they acted on those feelings in a selective fashion. Yet whether or not they practiced what they preached, it is significant that certain men of letters lauded a new model of behavior: the sentimental savant whose private life, and particularly his family life, testified to his virtue.[6]

## SOCIABILITY, SENTIMENT, AND FAMILY LOVE

The ideal of the married man of letters germinated slowly, in part because of the enduring appeal of bachelorhood. The idea that thinkers should surround themselves with like-minded men and scorn the burdens of marriage was an assumption inherited from the Middle Ages, when philosophers had prescribed celibacy as the surest path to intellectual excellence. That philosophers also tended to be men of the church only strengthened the link between celibacy and the life of the mind, for they vowed to remain unmarried as a way to keep their thoughts pure and their devotion to God unwavering. Marriage and the obligations of family life would distract the philosopher with worldly concerns, when he should be training his gaze on more universal targets.[7] The monastic tradition—the foundation of Western European intellectual culture—created a legacy of solitude and seclusion as the foundations of merit. This changed in the late medieval period with the advent of new funding sources for intellectual work. Philosophers became less reliant on church benefices but were left to their own devices when it came to managing their households. Confounded by the amount of work that household management entailed, they discovered they needed a woman to take care of them. Driven more often than not by practical concerns, philosophers began to marry. Yet even as medieval philosophers relied on their wives to look after them, they belittled their spouses' intelligence and disparaged their usefulness.[8]

Philosophy and family life had begun to coexist, but they made uneasy bedfellows. Bachelorhood remained the most desirable way of life. Indeed, philosophers' disdain for marriage endured long past the medieval period. Even when some Renaissance humanists and Protestant theologians encouraged philosophers to wed, they did not necessarily dismiss the attractions of bachelorhood.[9] The same dynamic appeared in the seventeenth century. Savants pursued increasingly social means to develop and display their ideas and yet still considered bachelorhood the ideal state.

Enlightenment philosophes, however, claimed to be unlike all philosophers who had come before them. Although they exaggerated the differences between themselves and their predecessors, new ideals of engagement and philosophical living did indeed emerge during the eighteenth century. A number of factors influenced intellectual self-fashioning during the Enlightenment, most especially ideas of sociability, sentiment, and civic virtue. These, in turn, shaped perceptions and representations of married intellectuals.

Scholars pinpoint the late seventeenth and early eighteenth centuries as the birth of a new concept: the invention of society as a stand-alone field of human existence.[10] "Society" had once referred to a select gathering of people, but the term came to evoke something grander and more abstract in the eighteenth century. It expanded into its modern definition as the arena of human relationships and interactions. Moreover, the number of references to "society" in published texts skyrocketed. Society elbowed aside Christian metaphysics and absolutist politics as the organizing principles of people's lives. To be socially virtuous, to be useful to society: these were the real markers of goodness. The reward for good behavior—the love and appreciation of one's peers—was likewise secular. Many people continued to practice Christianity, of course, and many hoped that they would enjoy a heavenly afterlife. But that far-off reward no longer stood on its own. More often than not, "society" and its needs served as motivating principles.[11]

As "society" came into its own, seventeenth- and eighteenth-century writers found themselves in the grips of a new obsession: sociability. Many were inspired by the writings of Anthony Ashley Cooper, 3rd Earl of Shaftesbury. Shaftesbury wished to show that human beings were naturally sociable creatures. "What Wretch is there," he asked, "what open violator of the Laws of Society, destroyer or ravager so great, who has not a Companion, or some particular Set, either of his own Kindred, or of such as he calls Friends, with whom he shares his Good, in whose Welfare he delights, and whose Joy he makes *his* Joy?"[12] Sociability was a good thing, as passions and attachments inclined individuals toward virtuous behavior. If people acted selfishly or unsociably, they would be exiled from society; this theoretically deterred them from immoral behavior.

Denis Diderot found these ideas intriguing, and he ruminated on them at length in his 1745 translation of Shaftesbury's work. Diderot proved himself an active translator, expanding some parts of Shaftesbury's philosophy, truncating others, and interjecting his own voice as he saw appropriate. For example, the dangers of hypothetical solitude in Shaftesbury's hands morphed into a critique of not-at-all hypothetical religious solitude after his translator was done.[13] Although Diderot tweaked Shaftesbury's ideas about personal happiness and social harmony, the idea of a natural human sociability resonated with him.

Diderot was not alone. Many French philosophes believed that humans were naturally sociable, and that nurturing sociability made one happier and more virtuous. César Chesneau Du Marsais, for one, insisted in the *Encyclopédie* that "man is not a monster who must live only in the abyss of the sea or in the depths of a forest: the very necessities of life make commerce with others necessary to him; and in whatever state he may find himself, his needs and well-being draw him to live in society."[14] A philosophy oriented toward personal happiness need not be selfish, these philosophes insisted. By enjoying themselves and reaping the pleasures of life in the world, individuals would become better members of society and would contribute to the happiness of others. This was the true aim of society. Not security, not contractual obligation, not even divine will: personal and collective happiness stood out as the *raison d'être* of life on earth. A sociable life was a life as nature—that eighteenth-century key-word for all that was virtuous and orderly—intended.

Sociability might have been natural, but it needed cultivating. And that cultivation started at home: the family was the first place where individuals learned how to be sociable (and, accordingly, how to be virtuous). Marriage developed one's social credentials, for women were integral to the pursuit of civility and sociability. Women were believed to be naturally skilled at correspondence and oral communication, and they would help civilize their husbands.[15]

Hence, for the philosophes, virtue no longer entailed isolation from society or abstinence from family life *à la* the regular religious: instead, seclusion represented a moral hazard. The family became the key to both happiness and morality, on a personal as well as a public scale.[16] The Abbé de Mably summed up this belief in succinct prose: "Whoever does not know how to be a husband, a father, a neighbor, or a friend will not know how to be a citizen. In the end, domestic morality [*moeurs*] determines public morality."[17] This idea only acquired greater significance during the French Revolution, when a virtuous home life became necessary for "making all households happy, developing good morals, [and] contributing to public happiness."[18]

Praise of domestic life only amplified as ideas of sensibility and sentimentalism caught on. Sensibility—defined in the eighteenth century as a faculty, the ability to perceive and respond to outside stimuli—had moral overtones.[19] A person with an appropriately delicate sensibility would have good morals; one with dull senses would be insensitive to the needs of others. It became highly fashionable to display one's sensitivity through seemingly spontaneous (and therefore authentic) displays of emotion, with crying, weeping, and fainting being especially popular choices. The enthusiasm for outward displays of emotion grew so strong that William Reddy has coined it a new "emotional regime."[20]

Literary culture helped amp up the affect. Celebrated texts such as Graffigny's 1747 *Lettres d'une Péruvienne*, Sedaine's 1765 *Le Philosophe sans le savoir,* Diderot's 1758 *Père de famille,* and Rousseau's 1761 *La Nou-velle Héloïse* belonged to the newly popular genres of the *drame bourgeois* and the sentimental novel. The *drame bourgeois* focused on the every-day lives of ordinary people, rather than heroic myths or tales of martial glory. With sweeping gestures, weeping, and melodramatic speeches, ac-tors hoped to move their audience to tears, a satisfying crescendo of emo-tion that greatly appealed in the age of sentiment. Sentimental novels had a similar effect. Eighteenth-century readers clamored for moving stories, sharing the joys and sorrows of the characters they encountered. As many novels were epistolary, readers felt like they were witnessing private dra-mas and glimpsing the inner lives of characters. Men and women wept as they read, and delighted to have such clear evidence of their virtue, sen-sitivity, and humanity.[21] Novels and plays alike were thus explicitly de-signed to evoke a heady emotional response from their readers. And what sorts of scenes best prompted virtuous emotions? Often these authors fo-cused on domestic life as dramatic, emotional, and of fundamental impor-tance.[22] Once again, all roads led back to the family home.

Sentimental love stepped off the stage as a wave of individuals claimed to practice a new kind of family life, one founded on affection, personal choice, and mutual devotion. Sentimental domesticity was well in vogue by the second half of the eighteenth century.[23] Ideals of love, companion-ship, and trust were hardly new; it is easy to find evidence of married cou-ples treating each other with respect and affection in any period. But the language of sentimentalized domesticity took on a new weight during the Enlightenment. A growing number of individuals chose to represent themselves as loving and intensely emotive fathers, mothers, or spouses; these representations reflect what Kate Retford has called "the increasing importance of proper performance of familial life as a sign of merit." Over-flowing and spontaneous expressions of family love became the latest fad in self-fashioning, a way to make one's merit and good morals legible to the public. In portraits, mothers and fathers no longer posed sedately with solemn children; instead, children wrapped themselves adoringly around their mothers' necks while affectionate fathers looked on.[24] Sentimental domesticity was not a universal ideal, as it was most appealing to urban and literate populations. But it was far from a fringe phenomenon.

As the appeals of family life waxed, those of celibacy waned. Celibacy had once heralded admirable characteristics such as self-restraint and reli-gious devotion, but philosophes associated it with despicable traits. Choos-ing to lock oneself away in a convent or monastery suggested antisocial tendencies and corrupted morals. Unmarried laity and clergy alike fell under attack as celibacy became linked with luxury and depopulation.[25] Isolation of any sort lost its allure in the eighteenth century and instead threatened to tear the social fabric apart.

Medical writers added a new layer to this debate: celibacy and solitude were unhealthy. Philosophers were not immune from such concerns. Doctors such as Samuel-Auguste-André-David Tissot (1728–1797) worried that the seclusion and celibacy traditionally associated with the contemplative life took a physical toll on savants. Tissot argued that intellectuals, like masturbators, spent too much time alone and squandered their vital juices on selfish pursuits, possibly resulting in sterility. Indeed, celibacy and isolation could result in all manner of nasty physical and psychological symptoms; the secluded philosopher could look forward to developing madness and hemorrhoids, among other afflictions.[26] The surest safeguard against such unpleasant ailments was emotional, sexual, and rooted in sensibility. Although it was possible to have too much of a good thing, loving emotions tended to strengthen body and mind. The same could be said of a healthy sex life. The medical message was clear: pursue social relationships, or face dire consequences.

Perhaps the strongest sign of marriage's social value was the many Catholic clerics who lobbied for their right to marry. Enlightened clerics, such as the Abbé Saury, Abbé Desforges, and Abbé de Saint-Pierre, published tracts like Desforges's 1758 *The Advantages of Marriage and How It is Necessary and Healthy for the Priests and Bishops of this Time to Wed a Christian Girl*. Marriage, these clerics insisted, would integrate priests more fully into the nation and would enhance their social utility. Rather than re-treading the same abstract advice, they would be able to draw on their own experiences when advising their flocks on marriage and fatherhood.[27] These clerics wanted to make themselves into living examples of their beliefs.

Philosophes likewise felt pressure to practice what they preached. They wrapped themselves in the mantle of public leadership and claimed they had a mandate to make the world a better place.[28] They held themselves up as moral exemplars, public leaders whose patriotism and social engagement distinguished them from frivolous *bel esprits*.[29] A pair of articles in the *Encyclopédie*—Du Marsais's "Philosophe" and Voltaire's "Man of Letters"—reveal philosophes as anxious to assert their sociability and public utility. Du Marsais exulted that "for [the philosophe], civil society is, as it were, a divinity on earth; he flatters it, he honors it by his probity, by an exact attention to his duties, and by a sincere desire not to be a useless or embarrassing member of it." This love—nay, worship—

of society made Enlightenment philosophes superior to those "insensitive sage[s] of the stoics," who, Du Marsais sneered, had been "nothing but phantom[s]."[30] Voltaire concurred that men of letters had reoriented themselves toward society and that, as a result, they were better thinkers. By dedicating themselves to eradicating public evils such as superstition and false knowledge, they "served the state."[31] This was their brand of patriotism. For eighteenth-century writers, patriotism connoted a universal emotion—a selfless connection to all those, known and unknown, of one's nation. To be a patriot was to love humanity and to devote one's life to the public good; to not be a patriot was to live a narrow, self-serving life. For these writers, public service was their patriotic duty, and philosophes could best help the public by providing good examples of virtuous sociability. Sociability trumped seclusion.

There was thus a great deal of cultural ferment in the eighteenth century. Sociability, sentiment, domesticity, and civic virtue were infused with fundamental importance and became watchwords for the virtuous man of letters. Philosophes meant to model these ideals, but soon realized there was more than one way to be sociable and civic-minded.

## SOCIABLE BUT SINGLE

Many men of letters rallied to the calls of sociability and civic virtue, but some remained unmarried. This sort of savant crops up frequently in the *éloges* (eulogies) written by Bernard le Bovier de Fontenelle and in Charles Perrault's *Les Hommes Illustres qui ont Paru en France pendant le XVII siècle* (1697). These texts and the many others like them aimed to inspire readers to emulate the inner virtues of French heroes. They highlighted personal goodness, not showy heroics or courtly suavity, as the definitive mark of greatness.[32] These sources held up savants as men worthy of imitation, and they tended to dwell on private virtues such as independence, modesty, and devotion to the public as evidence of greatness.[33] Eulogizers and biographers happily recounted exemplary achievements, but they devoted considerable space to mundane events and everyday virtues.

In crafting stories of savants intended to inspire the public, Perrault and Fontenelle did not find marital anecdotes compelling. Overall, they restricted themselves to observing if and to whom a savant was married, how many children they had, and if those children could boast of any outstanding accomplishments. These recitations tended to be brief. For example, when Fontenelle wrote Jean-Élie Leriget de La Faye's eulogy, he simply noted: "He left behind one son from his marriage with Marie le

Gras, a woman from a noble family . . . a respectable woman of virtue and merit."[34] Perrault often used similar phrasing.

They sometimes wrote expansively about family life, as in Fontenelle's eulogy for Jacques Ozanam: "He married a woman with very little property because he was touched by her sweetness, her modesty, and her virtue. . . . Neither his studies nor his occupation stopped him from enjoying, along with her and their children, the simple pleasures that Nature attaches to the roles of husband and father."[35] At other points, Fontenelle and Perrault discussed children at length, especially if they were deemed worthy heirs. That happy families appeared amid these early *éloges* suggests that the ideal of the affectionately married and productive scholar had traction in the late seventeenth and early eighteenth century. That such representations only cropped up in the occasional entry, however, suggests that it had not yet picked up momentum.

Even if certain intellectuals could marry and have children without sacrificing their careers, Fontenelle and Perrault suggested that family life and intellectual work were not always compatible. The phrasing of Fontenelle's entry for Ozanam—that "neither his studies nor his occupation stopped him from enjoying . . . the simple pleasures that Nature attaches to the roles of husband and father"—implied that this would not be the case for everyone. Moreover, in the entry for Nicolas Le Fevre, Perrault quoted approvingly the thinker's ideas about celibacy: "I would like . . . to be as absolute in all my good resolutions that I have made regarding the conduct of my life as I am in the decision to never marry."[36]

Although Fontenelle did not criticize savants who chose to wed and have children, he praised those who abstained from such behavior. The eulogy for Isaac Newton, for example, lauded him for having been "occupied with an important responsibility . . . which did not let him feel any void in his life, nor need for domestic society." The eulogy for Alexis Littre also praised his austere bachelorhood: "He always worked with an exemplary assiduity for the Académie. . . . He never saw a single stage production. . . . He never thought of marriage."[37] In Littre's eulogy, family life was grouped with "frivolous" activities such as attending the theater and was in this instance contrasted with "exemplary assiduity."

These texts reveal a certain ambivalence about domestic life. Fontenelle and Perrault celebrated successful marriages, but they expressed some concern about the distractions of family life. They lauded bachelor savants and implied that such thinkers demonstrated a stronger commitment to the life of the mind and to the public. Men of letters *could* marry, they seemed to suggest, but should do so with caution.

Perrault and Fontenelle did not advocate a hermetic existence for men of letters. They expected thinkers to be sociable and dedicated to the public good. They made clear, however, that savants need not marry to prove their sociability and intellectual worth. Instead, male friendship served as the foundation of civility and virtue. In his entry on the bachelor Le Fevre, Perrault noted that "he formed an intimate and close friendship with Pierre Pithou, one of the most learned, wise, and best men of his era. They lived together and passed many years in such agreeable and virtuous commerce." Le Fevre's friendship spoke volumes about his sociability and virtue. Likewise, Perrault's discussion of Sainte-Marthe made clear that family love was not essential to prove oneself "honneste": he was an accomplished savant who "joined to [his intellectual achievements] the qualities of an upstanding man. He was a good friend, full of zeal for his Country, and unbreakably loyal in his service to his Prince." Friendship again provided crucial evidence of personal virtue.[38]

Although male friendship had pride of place, other relationships merited glowing descriptions. Many savants, even those who remained unmarried, formed households. Fontenelle and Perrault described how these men cared for their servants, looked after their relatives, adopted their students as their children, and dispensed charity. Caring for their wives and biological children was far from the only way that savants could found households or demonstrate their virtue and sociability.

Fontenelle and Perrault thus operated under the assumption that the ideal man of letters would form emotional bonds and serve his fellow citizens in the Republic of Letters. That they infrequently discussed marital love suggests they did not find it essential proof of a thinker's capacity for love and loyalty. Instead, male friendship was the most significant iteration of intellectual sociability.

In imagining a homosocial intellectual world, Fontenelle and Perrault were part of a long early modern tradition. Social interactions—informal gatherings, academic meetings, and learned friendships—were the glue that bound the Republic of Letters together.[39] Many men of letters believed conversation and correspondence were essential to the development of ideas, and they thirsted for the knowledge that erudite letter writers could provide. They gathered in societies and academies, which would develop their ideas through commerce with like-minded individuals. Learned sociability helped form and maintain social bonds of all sorts, from deferential esteem to egalitarian comradeship. Friendship bound far-flung savants together and prevented unseemly and unproductive conflicts from erupting. Unfriendly actions—which could include being overly boastful, demanding, unappreciative, or unsociable—could result in expulsion from the Republic.[40]

For much of the early modern period, most savants remained ascetic bachelors or, as Anne Goldgar has called them, "literary saints."[41] Their civility and commitment to society was never called into question as a result of their bachelorhood. Many savants chose not to form families, with no harm done to their reputations.[42] Luminaries such as René Descartes, Michel de Montaigne, Robert Boyle, and Isaac Newton found a certain degree of alone time essential to their work, and they embraced bachelorhood as a way to ensure it.[43] There were notable exceptions, such as the Dacier family, but celibacy and solitude remained meaningful categories for masculine virtue and intellectual authority long after philosophers had begun to marry.

This tradition persisted into the eighteenth century, although it was informed by new ideas about sociability. Jean le Rond d'Alembert's 1753 "Essai sur la société des gens de lettres et des grands" described philosophical life as marked by independent contemplation and the ideal man of letters as devoted to universal interests.[44] Although the ever-sociable d'Alembert did not expect men of letters to live as hermits, he recommended they avoid certain social ties and financial burdens, including marriage. This would protect their intellectual integrity and give them the freedom to pursue a busy social schedule. Another such philosophe was Louis-Sébastien Mercier. In his *Bonheur de Gens de Lettres*, he described the ideal man of letters as a genius with time for quiet contemplation; marriage and children only got in the way.[45] These men of letters assumed that savants would socialize and form emotional bonds, and that these connections would enhance their intellectual work. Yet marriage was not part of that equation. It had little to do with scholarly and civic merit. For savants who embraced a more sentimental worldview, however, that was about to change.

## IMAGINING MARRIED MEN OF LETTERS

Even though men of letters did not have to marry, the eighteenth-century emphasis on sociability and civic virtue and their foundation in domestic spaces opened up new ways of thinking about philosophes and family life. Married philosophes could claim a certain moral authority by representing themselves as virtuous family men. This new form of intellectual self-fashioning complemented, rather than competed with, other trends, including the need for philosophers to be polite and public-minded.

Néricault Destouches's 1727 play *Le Philosophe Marié* (*The Married Philosopher*), which explored the connection between marriage and philosophy, offers a logical place to start looking for representations of married philosophers. The play, subtitled *The Husband Ashamed to Be One*, opened by recycling tropes about the incompatibility of intellectual work and family life, such as the idea that a wife would be an unwelcome distraction. In an early scene, the protagonist, Ariste, scolded his wife, Mélite, when she appeared in his office: "You are as dear to me as you could possibly be / but must you be here in my study?" Ariste did not just consider his wife a sometime hindrance: he worried that marriage would be his social and intellectual undoing. Before his nuptials, Ariste had insisted that a truly dedicated philosopher would remain celibate and therefore wholly devoted to his craft. What would his peers say, he worried, when they learned that he—of all people—had taken a wife? When his father queried, "Why are you so ashamed of your marriage?," he responded: "Why? Because it will bring great ridicule upon me / All those whom I have mocked will mock me in turn."[46] Ariste's fears were further compounded by his concern that his domineering uncle would disinherit him if he learned that Ariste had married without his consent. Yet in the end, Ariste's fortune was secured and his love for his wife acknowledged without dire consequences. By highlighting Ariste's loving marriage as testament to his good character, Destouches suggested that philosophes could marry, and that marriage could enhance their happiness and virtue. The playwright did not imply that all philosophers must marry, but he did model a new way of life for savants.[47]

Some men of letters found this vision of family life seductive. Why should they deny themselves the pleasure and stability that marriage offered? Would they not better prove their public virtue by choosing to wed, rather than selfishly remaining unattached? Sometimes they even expressed such feelings in verse, as in Doigni Du Ponceau's 1773 "Letter to a Celibate Man of Letters." He mocked thinkers who feared marriage: "Marriage makes you shudder! O somber Celibate! / You disdain the titles of Husband and Father / Cold indifference dries out your heart, / And yet you think that by loving nothing, you ensure your happiness!" The poem, to which the Académie Française accorded an honorable mention, then counseled that "the true Sage consoles and serves Humanity, / Devoting his whole being to Society / and he does not push away his cherished Companion / Who helps him bear the burdens of life."[48] Replete with sentimental touches and many an exclamation mark, the poem stressed that married savants had a clear moral edge over those who never wed.

The *éloges* written for members of the Académie des Sciences at the end of the eighteenth century, as well as collective biographies of illustrious men, further speak to the rise of sentimental domesticity. These authors continued to valorize male friendship, just as their predecessors Perrault and Fontenelle had done, but they showed a growing confidence that companionate marriage also ensured a savant's happiness, virtue, and productivity. This new preoccupation with family love is evident in the eulogies written by the Marquis de Condorcet (secretary of the Académie des Sciences from 1777 to 1793). Although Condorcet did not take over where Fontenelle had stopped—Jean-Jacques Dortous de Mairan and Jean-Paul Grandjean de Fouchy came in between their tenures as secretary of the Académie—Fontenelle and Condorcet make for a useful pairing because they bookended the eighteenth century and because each wrote a large number of eulogies.

Read alongside Fontenelle's *éloges*, Condorcet's essays featured much lengthier discussions of family life and especially marital love. Indeed, sprinkled liberally with emotional outbursts, Condorcet's *éloges* were models of *sensibilité*.[49] He lavished attention on domestic anecdotes as well as stories of friendship; these proved a savant's virtue, sociability, and civility. He waxed enthusiastic about how deeply savants had loved their wives, how much time and effort they had poured into their children's educations, and how beloved they had been by kith and kin. Even if other virtues might have been more "brilliant," it was "much more touching" that savants developed close bonds with their spouses and children. Sentimental private life—being a "good friend, good son, good husband, good father"—was so admirable, in fact, that it could partially compensate for other character defects, such as a "love of dissipation."[50]

Exclamations about the joys of marriage and fatherhood were common in Condorcet's eulogies. In eulogizing Jean-Baptiste Bucquet, for example, Condorcet enthused that he "was married; he wed one of his relatives, whom he loved and who loved him. . . . He was sure that he had found in his wife a tender friend, and that she would be a good mother to his children; he felt himself necessary to her happiness, just as she was necessary to his own." Likewise, in his eulogy for the Marquis de Paulmy, Condorcet exclaimed, "His life passed in the bosom of his family was sweet and peaceful; an exacting probity, a noble and disinterested conduct in his particular affairs, earned him the respect of those who had personal relations with him."[51] Condorcet's language resembled that used in tracts advocating romantic marriage and legalizing divorce (which would have made it possible to escape loveless marriages). For example, the pamphlet writer

Cailly effused that the ideal "family [was] united and propagated by love. All are in harmony, peace reigns in this household. Father, mother, children: all love each other. The conjugal tenderness that the spouses feel in their hearts has not changed at all. . . . Their children, brought up according to the purest of principles, have made [virtue] a habit."[52] Furthermore, Condorcet did not represent marriage as a distraction; instead, intellectual accomplishment and marital love fed into each other. He imagined that for Madame Buffon, "each of her husband's new works, each new palm attached to his glory was for her a source of pleasure of the sweetest sort."[53]

Older traditions persisted alongside new ones. Condorcet did not claim that married savants lived lives free of complication or inconvenience. In his *éloge* for Bucquet, for example, he noted that the savant faced financial pressures in providing for his family and worked himself raw (although he was motivated, perhaps even more so, by his own love of science and desire to win the approval of his colleagues).[54] Furthermore, Condorcet praised many sorts of relationships beyond those of marriage and biological parenthood. His discussions of male friendship dwelled on themes of love and kindness, and he also wrote drawn-out descriptions of savants taking in relatives and loving their students as their own children. Although happy marriages provided much grist for his mill, so too did all manner of friendships and household arrangements.

The ideals of affectionate marriage were also found in biographical collections such as Louis-Pierre Manuel's *L'année françoise,* a series of essays that celebrated one illustrious Frenchman for each day of the year. This curious text was cobbled together from various other sources, including eulogies written by Fontenelle and Condorcet. Some essays continued to depict the ideal man of letters as single. Other essays, lifted from various sources, dwelled on the affective bonds between men of letters and their families. That Manuel held up bachelors *and* husbands as models underscores the coexistence of married and unmarried savants as admirable figures. Many essays were unoriginal in content, but Manuel obviously considered the ideals they represented to be worthy of imitation. Moreover, he heavily excised the original texts to focus on personal virtues, including friendship and family love; this focus is significant.

Although friendship remained important, Manuel's essays stressed the significance of family life as a source of virtue. He noted that Théodore Tronchin, the famous Genevan doctor, was "received and sought after in the most brilliant circles but he was never happier than with his family, the picture of republican simplicity." Manuel also recounted various stories of the playwright Racine's loving family life. Their inclusion was not accidental: Manuel claimed that "nothing elevates a great man like these domestic anecdotes."[55] In his introduction and in his sentimental descriptions of marriage and family life, Manuel's *Année* suggested that the married philosopher could be very virtuous indeed.

These late eighteenth-century eulogies and biographies demonstrate new ways of imagining thinkers' private lives and virtues. In their earlier texts, Fontenelle and Perrault had primarily discussed family life as proof of a savant's diligence and commitment to duty, with the occasional nod to a particularly successful marriage or especially devoted parenting. Both valued virtue and civility, but these qualities were sufficiently demonstrated by an individual's friendships, household, and institutional affiliations. Condorcet and Manuel shared Fontenelle and Perrault's general assumptions about virtue and social utility, but they relied on marriage as well as friendship as evidence of such.[56] They stressed the compatibility of family life and intellectual work, as when Manuel described a savant's life after marriage thusly: "Married, he continued his simple, quiet life, but now with a singular happiness as marriage made his household more pleasant. His studies profited from it."[57] Late eighteenth-century writers discussing family love were also more likely than earlier authors to infuse their writing with poetic declarations of the joys of romantic, paternal, and companionate love. Their effusive texts were much closer in form to the sentimental novels of Rousseau and Richardson.

Eulogizers and biographers heaped praise on social relationships of all sorts, not simply those of marriage and biological paternity, making clear that men of letters had more than one path to choose. They could live as bachelors, or they could marry and revel in the delights of family life. Although Condorcet lauded those savants who enjoyed loving and productive marriages, he did not criticize those who chose to remain unmarried. As many savants continued to gravitate toward bachelorhood, Condorcet wrote several eulogies of unmarried men. Like their married compatriots, Condorcet made clear that these individuals were virtuous, beloved by their friends and family, and committed to their studies.

Condorcet's eulogies showed that both marriage and bachelorhood remained viable options; new representations of married savants were compelling yet did not eliminate other models of living. He did, however, reject the idea that bachelorhood was the smarter, safer, and more laudable choice. "One finds, among bachelors and married men, men of equal genius," he wrote. "It would thus be unjust to reprimand a man of letters for living in either of these states."[58] He removed the stigma attached to married philosophers and insisted that their marriages and paternal affections offered compelling testimony to their character.

What was new in these texts was the language that savants used to describe their peers. The goal remained the same that it had been for Fontenelle and Perrault—to represent savants as learned but engaged, virtuous, and devoted to useful productivity—but the method was different. The language of feeling had become an idiom through which intellectual virtue was communicated. New ideas about sociability, civic virtue, and sentimentalism clearly informed Condorcet's and Manuel's understandings of the ideal man of letters.

Condorcet's ecumenical approach appealed even to those who embraced bachelorhood. For example, in *L'Homme de Lettres* (1764), Jean-Jacques Garnier posed the question "should a man of letters marry?" and ultimately answered in the negative; solitude and independence were essential to the life of the mind. Yet, at the same time, he acknowledged that Nature designed all men—even men of letters—to marry and reproduce. He conceded that marriage could prove an asset for a man of letters if he exercised care in choosing a suitable spouse, if she was "another Hypatia."[59] Hypatia had been an astronomer in fourth-century Egypt and was regarded as one of the most brilliant and accomplished women of all time. Although Hypatia herself had been famously reluctant to marry, Garnier singled her out as an ideal wife for a savant. In other words, he imagined that the only suitable wife for a philosophe would be a brilliant woman dedicated to intellectual pursuits. Presumably, the two gifted individuals would collaborate effectively, making theirs a union of the mind as well as the heart. In stressing that marriage could work for some men of letters, therefore, Garnier acknowledged the potential benefits of the new ideal of companionate intellectual marriage.

But what of a different sort of savant? One interested in mistresses, but not marriage? The astronomer Jérôme Lalande declined to marry and rationalized his decision by claiming: "I have loved women very much; I still do . . . but my passion for them is rational: it has never jeopardized my fortune nor my studies."[60] Yet the same savant, in his "Moral Testament," stressed how deeply he loved his illegitimate daughter and her family. When he wrote of them, as opposed to matrimony, he hewed more closely to sentimental norms: "My sensibility makes me prone to tears; this is especially brought about by my attachment to my family, which is one of my most cherished duties."[61] Family love was crucial for Lalande's personal happiness, and his clear attachment to his kin—evidenced here, as was typically the case, by his tears—demonstrated his sensibility and virtue. He may have rejected marriage as unappealing, but he cobbled together a sentimentalized family of his own.

That Lalande found the figure of the sentimental savant meritorious is clear from the way he discussed other men. When painting a picture of other men's married lives, he dipped his brush deeply in rosy hues of romantic love. In a eulogy written for his friend, the naturalist Philibert Commerson (d. 1773), Lalande waxed effusively about the love shared by Monsieur and Madame Commerson and the tragic circumstances that had befallen them: "This sweet and charming union kept him occupied for two years . . . but the birth of [their] child cost the mother her life." Commerson was devastated by his wife's death, writing to a fellow widower that "I have lost, as you have, the most tender and virtuous of Spouses, and I only exist today through the memory of having belonged to her."[62] The naturalist carried on with his work and, in the course of his travels, discovered a new plant species that resembled two hearts that had grown together. Commerson used this opportunity to commemorate the other half of his own heart, his much-mourned wife, by naming the plant the *Pulcheria commersonia*.[63] Lalande represented Commerson as sensitive and devoted to the memory of his spouse, and stressed that he had combined his intellectual and romantic passions. Lalande found sentimental language appealing, even though he structured his life outside the lines of the normative ideal.

The same was true of Voltaire and Émilie Du Châtelet. The couple first met in 1733 but never married, for Du Châtelet already had a husband. Theirs was an aristocratic love affair with an intellectual twist. They fêted their romance as an ideal blend of love and learning. Of her, Voltaire wrote: "Consider . . . what infinite attachment I must feel towards a person in whom I find the reasons to forget everyone else, in whose company I learn new things every day, and to whom I owe everything."[64] Du Châtelet's husband proved remarkably tolerant of her relationship with Voltaire, and so the two were able to live together at Cirey, her estate in Champagne. They made the estate into something more than just a private residence. They imagined a place where they would each have space to conduct experiments, contemplate, and write.[65] In demonstrating how nicely love and intellect could fit hand in hand, the couple may have prefigured subsequent attempts at intellectual companionate marriage, even though they were never married to each other. Du Châtelet and Voltaire's extramarital relationship modeled the intellectual benefits that romantic love could proffer, albeit outside the institution of marriage.

The idea that romantic love and affectionate family life were virtuous and useful had thus become widespread, and could even be promoted by the occasional bachelor. Some savants remained single and could be celebrated as such, but the figure of the happily married savant had become a powerful normative ideal. In accord with new ideals of sentimentalism, sociability, and civic virtue, married men of letters drew on a sentimentalized language of family life to depict themselves as admirable and trustworthy figures.

## SENTIMENTAL SAVANTS: THEIR LIVES AND LETTERS

The ideal of the sentimental savant gained purchase with various men of letters, and it became a part of their self-representations. The examples of two married philosophes, Claude-Adrien Helvétius and Denis Diderot, illustrate how two very different men of letters fashioned themselves as sentimental savants. Helvétius married Anne-Catherine de Ligneville in 1751, and shortly after his marriage, he made direct reference to Destouches's play in a letter to a friend: "I am the married philosophe. If you knew how this confession costs me and how shameful I feel, you would excuse my foolishness, for I still have enough sense to know that marriage is ridiculous."[66] He wrote a similar letter a few weeks later in which he playfully confessed, "Yes, truly, Monsieur, I have married. I have sinned against philosophy and I demand absolution from you, first among philosophes. It is the foolishness of love, of which I cannot repent, and I can do nothing better in this respect than to die unrepentant."[67] The language here calls to mind that of medieval intellectuals, who had viewed celibacy as essential so that scholars might "marry" philosophy.

Yet Helvétius's discussion of his marriage also departed from medieval norms in significant ways. Medieval scholars had seen marriage as a pragmatic means to obtain household help, but they had downplayed their attachment to their wives. Helvétius, however, claimed that his decision to wed was motivated by love and exulted to his wife, "I am the happiest man in the world because I am loved by such a lovable woman."[68] He later wrote: "Know that I will love you to the grave and that my last breath will be for you."[69] The initial flame of love only grew stronger. "If you were here with my children and two or three of my friends, I would find myself happier here than anywhere else. But you, especially, are the source of my happiness. I love you more and more, I miss you as a friend and as a lover." He claimed that his feelings for her were so strong they bordered on insanity: "As for me, I love you to the point of madness, and I will repeat it with intense pleasure. I would like to tell you what I feel and I imagine that in repeating 'I love you' I will make you understand me."[70]

In such letters, Helvétius described his wife as the love of his life, and he wrote about his feelings in rhapsodic terms: "You know how much I love you, how dear you are to me, how touched I am by all the signs of friendship you have given me, how sure I am of your tenderness and of your love, and how this certainty fills my life with happiness; it flows through my veins like milk."[71] These were not isolated outbursts, as his letters to his wife were peppered with endearments: *ma chère amie, ma chère enfant, je t'aime de toute mon ame et de tout mon coeur, je t'adore.* Nor was this a strictly private matter. Eighteenth-century biographers used his love for his wife and children as proof that he was a kind and sensitive man.[72]

Helvétius also represented his marriage as a contented intellectual partnership. In a letter to Charles Palissot de Montenoy, Helvétius wrote: "My wife . . . shares my opinion of your work. We are like citizens or amateurs of letters enchanted by your talents."[73] The letter framed all praise as emanating from both of them, implying that the couple had read Montenoy's verses together and developed a shared opinion. In this letter and in others, Helvétius described himself and his wife as an intellectual unit. He presented his marriage as both fulfilling and productive. That Helvétius enjoyed a comfortable income and aristocratic status further enabled his pursuit of sentimental norms: he had the time and income to support a family.

Helvétius's loving marriage was only part of the story, however. Even as he declared his overflowing love for his wife, he sought the company of prostitutes.[74] It was not unusual for an aristocratic man to pursue sexual pleasure in this way, and Helvétius's letters give no indication that he considered such behavior out of the ordinary or even hypocritical. Instead, he seems to have had it both ways: aristocratic *galanterie* and sentimental domesticity. The one did not necessarily replace the other. Helvétius seems to have been sincere but selective in adopting sentimental norms.

Like Helvétius, Denis Diderot filtered his letters through a sentimental lens, but he placed greater emphasis on his role as a father. Ever strapped for cash, in 1765 he asked his friend and fellow philosophe Melchior Grimm to write a letter to General Betzki, an associate of Catherine II, to determine if the empress would support him as his patron. In this letter—presumably written in consultation with Diderot—Grimm wrote, "The philosophe Diderot, after thirty years of literary employment, finds it necessary to do away with his library so as to provide for the education of his only daughter."[75] Through Grimm, Diderot presented himself as both father and philosophe. Rather than one excluding the other, his literary career would fund his daughter Angélique's education.[76] His middling social status may have made it difficult to support his family and live as a philosophe, but he found a way to compensate.[77]

Diderot's fondness for his daughter was not unusual, but his portrayal of himself as a father was a concerted attempt to integrate the roles of philosophe and father. It constitutes a rethinking of the relationship between intellectual work and family life, a new way of representing the two as mutually beneficial. His rhetorical strategy proved successful. Catherine II, according to General Betzki, was touched by Diderot's good intentions. Betzki wrote, "Her compassionate heart could not look without emotion on this philosophe, so famous in the Republic of Letters, who finds himself in the situation of sacrificing to paternal tenderness the object of his delights, the source of his labors, and the companions of his spare time."[78] Agreeing to act as his patron, she purchased his library at the price he proposed and granted him a pension.

The self-representation contained in this letter and its reception speak to a new kind of philosophical ideal, one in which men could pursue the roles of scholar and father with mutual benefit. Although Catherine's letter acknowledged the tension between philosophical and family life by describing Diderot's "sacrifice" of his work for his "paternal tenderness," Grimm's letter did not. By representing himself as both father and philosophe, Diderot helped create a new intellectual ideal, a man of letters who was able to revel in the particular interests of his family and to contemplate the universal concerns of philosophy.

Other correspondence featured letters that positively oozed sentiment. In a letter written during their courtship, J. B. A. Suard assured Amélie Panckouke (sister of the famed publisher) that "you must well persuade yourself that I love you with my whole heart, that nothing in the world can alter this sentiment of mine, and that I will make my greatest happiness that of assuring your own."[79] Pierre Choderlos de Laclos wrote to his wife Marie-Soulange Duperré that "this reciprocal tenderness which reigns in my family is a treasure which increases by sharing it, and which one can enjoy without changing it; precious qualities that other treasures do not have."[80] It was only reasonable that he should love her, he wrote, for "you have certain qualities which justify my love for you. . . . Adorable mistress, excellent wife, and tender mother: such is the list in just a few words. . . . I congratulate myself every day for having found you."[81]

These letters show that late eighteenth-century men of letters represented their marriages as loving, affectionate, and virtuous, and that they wanted to portray themselves as the kind of men who would attract and appreciate romantic feelings. Eighteenth-century correspondence was often shared with close friends and family, and so when savants wrote sentimental letters they did not do so for an audience of one. Instead, correspondence was a way for letter writers to project a particular image of themselves. It is thus telling that savants wrote letters that seemed as if they had been ripped from the pages of a sentimental novel. They had become fluent in the language of feeling and used it to represent themselves as loving and virtuous members of society.

In fact, some savants deliberately imitated Rousseau, master of the genre. When Madame Laclos compared her husband's letters to those of the fictional lovers in Rousseau's *La Nouvelle Héloïse*, Laclos was delighted. "You find that Rousseau and I write in similar ways! You do me much honor in saying so," he exclaimed, "for he has written of nearly everything that you have inspired and continue to inspire in me." Laclos went still further in writing that "[Rousseau] and I are possibly the only ones capable of speaking in the language of your heart; you know so well how to hear and appreciate our words."[82]

Sometimes, letter writers directly lifted lines from Rousseau's pages. As Laclos wrote, "I would have said with Rousseau and like Saint-Preux: 'Julie, Julie, our hearts will never be deaf to the other's.'"[83] Some years later, when he heard that their youngest child Charles was ill, Laclos again used Rousseau as a touchstone: "'What a cruel state absence is, in which one can only see another in the past, and in which the present ceases to exist.' Never, my dear friend, have I felt the truth of that phrase of Rousseau's more than I did when I received the news you sent about our poor Charles and his illness."[84] Using Rousseau as an emotional template continued into the nineteenth century, as illustrated by the savant André-Marie Ampère's courtship journal. He wrote: "I ate a cherry that she had dropped; I kissed a rose that she had sniffed; while out for a walk, I twice offered her my hand to help her step over things underfoot."[85] The line mimicked a speech by Saint-Preux in *La Nouvelle Héloïse*: "I saw nothing unless her hand had touched it; I kissed the flowers where her feet had trod; I breathed along with the roses the air that she had breathed."[86] These savants were among legions of readers who modeled their lives after sentimental novels.[87]

Gone were tropes of domestic distractions; these thinkers focused instead on portraying their marriages as happy and fulfilling. The importance of nuptial ties is further underscored by the complaints of bachelor savants who longed to be married, like the young Marquis de Condorcet.

As an unmarried man, he lamented that "my solitude afflicts me because it has made me aware, for some time now, that I no longer have the power to fill the voids in my life with work."[88] When he finally wed Sophie de Grouchy in 1786, he extolled the benefits of companionate marriage and delighted that he had found such an intelligent, vivacious, and beautiful wife.

Correspondence did not straightforwardly reflect lived reality. Anxious to impress their peers, mindful of posterity, and coached by various letter-writing manuals, eighteenth-century letter writers crafted their letters with care. Accordingly, letters did not include everything about a marriage; infidelities and arguments did not generally appear on these pages.[89]

Yet although love letters do not offer a clear window into the actual experiences of intellectual couples, they have their merits as historical sources. These correspondents represented themselves, their feelings, and their marriages in similar ways, which suggests the development of a new romantic ideal. Instead of complaining about the distractions of family life, late eighteenth-century savants represented their domestic lives as intellectually productive. Whereas their predecessors had warned that wives were likely to intrude on a man's study and tear him away from his work, these savants portrayed their spouses as their intellectual partners (albeit unequal ones). As the next chapter will show, they even trained their wives as collaborators, encouraging them to study foreign languages, mathematics, natural philosophy, and astronomy. Evolving representations of wives from shrewish distractions to loving assistants were undoubtedly driven by changing perceptions of gender and femininity. Many a writer in eighteenth-century France depicted woman as man's perfect complement: the passive helpmate to the active patriarch, the emotional caregiver to his rational provider, and so on.[90] Such ideas may have encouraged male savants to take advantage of their wives as capable assistants.[91]

Perhaps the most iconic of these affectionate and productive marriages was that of Antoine and Marie-Anne Lavoisier. Antoine enjoyed a healthy income courtesy of his position as a tax farmer for the General Farm, and he invested a sizable fortune in his scientific interests. Eventually, he gained entry to the Académie des Sciences and became famous for his experiments with the composition of air, and particularly with oxygen. He married Marie Paulze, the daughter of one of Lavoisier's colleagues in the General Farm.

Antoine and Marie-Anne became a celebrated model of companionate marriage. Contemporaries seemed particularly impressed by how well matched the two were intellectually, as many commented on Madame

Lavoisier's contributions to her husband's work. The Portuguese thinker Jean de Magalhaens called Marie-Anne an "épouse philosophique," while the English writer Arthur Young approvingly described her as "a lively, sensible, scientific lady."[92] The language used to describe their relationship was most often that of friendship, solace, and general contentedness. Even in dramatic periods of upheaval, these remained the basic motifs of their marriage. In Antoine's last letter to his wife—written in a revolutionary prison as he awaited trial for "crimes" committed while he had been a tax farmer—he wrote that he would die without regrets. "I have had a long and successful career," he wrote, "and have enjoyed a happy existence ever since I can remember. You have contributed and continue to contribute to that happiness every day by the signs of affection you show me. I shall leave behind me memories of esteem and consideration. Thus, my task is accomplished."[93]

The most spectacular—and spectacularly expensive—example of the Lavoisiers' self-representation is the portrait they commissioned from Jacques-Louis David in 1788 (figure 1.1). As Marie-Anne was a student of David's, she presumably took an active role in commissioning the portrait and shaping her and her husband's image. The composition features Antoine seated at a table covered in sumptuous red velvet and surrounded by scientific instruments; he holds a pen to the pages of his *Traité elémentaire de chimie*. Marie-Anne, the dominant figure in the painting, leans over his shoulder. Antoine looks back at his wife, perhaps to show her what he is working on, while she looks directly into the viewer's eyes. In this peaceful domestic and scientific scene, they seem to be working together. The portrait signals that the Lavoisiers wished to be viewed as partners, for David did not portray Antoine in the traditional pose of a solitary thinker. Instead, he depicted the Lavoisiers' life as a harmonious blend of public and private, affect and intellect, male and female.[94]

Furthermore, the painting suggests that theirs was a partnership based on complementary forms of labor. In David's portrait, Marie-Anne has paused over her husband's shoulder for just a moment; her work station—a drawing board where she was presumably sketching illustrations for the *Traité elémentaire*—is located across the room from her husband's desk. David portrayed them as working together but also separately. This division of labor is reflected in the way in which Antoine looks toward his wife but she looks toward the viewer. By fixing Marie-Anne's gaze in this manner, David evoked her role as the family hostess and public advocate of her husband's theories.[95]

Fig. 1.1. Jacques-Louis David, *Antoine-Laurent Lavoisier and His Wife*. 1788.
Oil on canvas, 102 ¼ × 76 ⅝ in. (259.7 × 194.6 cm). Metropolitan Museum of
Art, Purchase, Mr. and Mrs. Charles Wrightsman Gift, in honor of Everett
Fahy, 1977, 1997.10. Image copyright © The Metropolitan Museum of Art.

The image was practically a poster for the intellectual benefits of companionate marriage. Antoine and Marie-Anne came across as an ideal couple: in love, but not manically so; absorbed in their tasks but still attentive to each other and their surroundings; working for the good of the family and the good of the public. In short, the two represented a perfectly complementary couple. That was no accident, as complementarity between the sexes was seen as the foundation of companionate marriage. But theirs was a complementarity of a particular sort. Marie-Anne's performance of the dutiful and loving wife did not consist of caring for their children, as was the case with many portraits of the time. This was a more intellectual spin on complementary marriage, with Marie-Anne fulfilling the wife's role of loving helpmate by assisting her husband with his scientific work. The portrait thus makes clear the benefits companionate marriage could hold. Even if the reality did not quite match representation—Marie-Anne Lavoisier was, at the time the portrait was painted, involved in a romantic liaison with Samuel-Pierre Du Pont de Nemours—the fact that the Lavoisiers invested so much in this image suggests they considered it and the story it told vital for their continued social and scientific prominence. Depicting their marriage as happy, harmonious, and collaborative was a way for them to make a positive public impression.

The story the Lavoisiers were selling was coveted by many. Those whose marriages fell short of this ideal, like Denis Diderot, felt the loss keenly. Diderot repeatedly complained that his wife failed to support his work and that she was slow to attend productions of his plays. His disappointment speaks to his expectation that a good wife would have helped with, not hindered, her husband's work and public reputation.[96]

Such a lifestyle required a certain amount of capital, and the men of letters discussed here—especially Condorcet, Lavoisier, and Helvétius—were of comfortable backgrounds. They could afford to pursue family life and philosophy at the same time. Finances clearly mattered when savants considered marriage, as it did for all social groups.[97] Yet just as important were changing ideas of marriage, virtue, and social utility. Social status may have helped determine why some men of letters married and why others did not, but so too did their understanding of the intellectual and emotional value of family life.

The correspondence, images, and commemorations of savants thus show that the man of letters as family man became a widely recognized figure in the social imaginary by the end of the eighteenth century. Happily married men were fêted as virtuous and productive. In the context of debates about sociability, civility, sentiment, marriage, and gender, eighteenth-century thinkers forged a new normative ideal that celebrated intellectual companionate and collaborative marriage. Male savants began to portray intellectual life as enhanced by the experience of domestic affect. They represented themselves as a new type of public man, one who melded the particular interests of family life and the universal concerns of philosophy. Family life and philosophy became more than just compatible: they became mutually beneficial.

## CONCLUSION

Philosophers had been permitted to marry since the late medieval period, but intellectual ideals were slower to change. Depictions of ideal philosophers continued to valorize independence, celibacy, and seclusion long after the Middle Ages. But in the eighteenth century, new ideas about sociability, sentiment, and civic virtue encouraged men of letters to use the newly fashionable language of feeling. By stressing their affective ties and social connections, they could represent themselves as moral figures worthy of emulation. A variation on this theme was the man of letters who was also a family man. By emphasizing their happy domestic situations, married men of letters represented themselves as men of feeling. They waxed poetic about the joys of family life, pontificated on the social benefits of marriage and fatherhood, and used their loving family lives as evidence of their good character and public utility. The ideal of the sentimental savant had come into its own.

# CHAPTER TWO

# Working Together

M en of letters hoped to reap more than rhetorical benefits from their families. They also expected their wives and children to perform all manner of useful labor, including observing, calculating, illustrating, translating, and publicizing. No tidy separation existed between home and work: families of savants lived surrounded by papers, books, scientific instruments, and specimens. Households had long been integral to the making of knowledge, with kitchens and drawing rooms doubling as scientific spaces.[1] Families had often labored together, in intellectual as well as other contexts, but eighteenth-century men of letters put a new spin on this old model. They sentimentalized the family workshop, using "love" and "devotion" as keywords to describe the work done by their wives and children. They doubled down on affective rhetoric by portraying themselves as productive men of feeling in charge of loving and learned households. To bring this vision to life, they needed to do more than represent themselves as sentimental figures; they needed to bring their wives and children into the picture as well.

Representing family love and intellectual work as happy bedfellows shaped ideals of feminine intellectual merit. A domestic ideal of the *femme savante* shifted into focus: the learned and loving wife or daughter who devoted her considerable talents to collaborating with her husband or father. This idealized woman was not torn between family duty and intellectual work, but rather excelled at both. She pursued a "double day": a full slate of domestic duties along with a full day of intellectual work. Men of letters represented their female family members as exemplifying this ideal. They recognized their wives and daughters as active collaborators and stressed their unswerving devotion to their families.

By portraying their wives and daughters as loving and learned, men of letters moved the needle on representations of *femmes savantes*. They challenged the idea that philosophizing or scientific women were inherently odd. They implicitly responded to critics of women savants who claimed that women were too emotional or narrow-minded for serious intellectual work.[2] In contrast to a vision of learned women as masculine, unnatural, and untalented, sentimental savants held up an image of female collaborators as wholly feminine, nurturing, and skilled. To a certain extent, this new representation opened up opportunities for women to collaborate on intellectual matters without fear of ridicule. But in the end, the idea of the learned woman diligently and brilliantly assisting her husband remained a fundamentally male-centered vision. Although male savants often credited their wives for their work and praised them for their talent, they did so within a patriarchal framework that preserved men as the head of the household and the head of all research projects. Sometimes, the more things change, the more they stay the same.

When male savants conjured up an image of the perfect wife, they did not dream of a shrinking violet who would stay out of her husband's way. Instead, they imagined a more active household presence. Desirable characteristics in a wife—that she be a skilled manager of the household, devoted to her family, and tireless in her duties—paralleled qualities associated with intellectual merit such as passion for learning, attention to detail, and physical stamina. The perfect partner was learned, capable, modest, and loving, with emotional warmth and a sharp intellect.

Contemporary readers may find such patriarchal clichés tiresome. Indeed, reality was certainly more complicated than glossy representations allowed. Wives and children may have occasionally resented the power wielded over them; perhaps squabbles erupted. Perhaps the talented, ambitious, and seemingly indefatigable women discussed here might be better known to history if they had worked independently. All of this is certainly true, but eighteenth-century men and women still found the idealized family workshop appealing. They perceived great social and emotional value in a household founded by a man of feeling, supported by a learned woman, and sustained by the work of affectionate and well-trained children.

This chapter considers both sides of the coin: the work done by spouses and children and the way such work was represented in correspondence and publications. Married men of letters relied on their wives and children to help them with a range of projects. They then represented this work as proof of their families' love and devotion, a representation that reflected well on the male head of the household. This was all part of the ideal of the sentimental savant, the imagined next step for a learned man who immersed himself in domestic life. Family work had a concrete impact on the making of knowledge and was, at the same time, folded into the ongoing reimagination of the ideal man of letters.

## HOUSEHOLD MANAGEMENT

Family members had long contributed to life in the Republic of Letters by preparing meals, cleaning house, and balancing accounts. This was usually women's work, and was most often done by a wife. Necessary though it was, housekeeping generally went uncelebrated. Although intellectual historians often gloss over such work, eighteenth-century men knew to look for a savvy manager when choosing a wife.[3] Delegating household tasks to family members allowed thinkers to pursue their work without having to worry about mundane details.[4] Men of letters had to maintain the illusion of independence, disinterestedness, and dedication to their genius; it would not do for them to be distracted by their accounts.

Consider the Baron de Montesquieu, celebrated author of *The Spirit of the Laws* and *The Persian Letters*. His marriage to Jeanne Lartigue lacked warmth, yet his letters show that she enabled his work. Her dowry of 100,000 *livres* injected much-needed cash into his finances and allowed him to devote his energies to subjects other than his bills. She assumed much responsibility for their home and lands.[5] His duties as a magistrate frequently pulled him away from their estate, as did his intellectual interests and busy social calendar. The burden of running his properties fell to his wife. The Baronne de Montesquieu had a dab hand when it came to finances and her husband relied on her accounting.[6] Just as importantly, she acted as his eyes and ears when he was away from the province on business.[7] A wife like the Baronne de Montesquieu was an asset. Her efficiency granted her husband the freedom to travel and permitted him to devote himself to his work and writing.

Another case in point is the household of André-Marie and Julie Ampère. André-Marie is best known for his work on electromagnetism and the unit of measurement for electrical current that bears his name (colloquially known as the amp). Like his earlier predecessors, he benefited from his wife's financial stewardship, even though he lived apart from her for most of their marriage. Julie Ampère bore near-complete responsibility for running their Lyon household and raising their son. This allowed Ampère to work in Bourg as a professor of mathematics. Julie's letters indicate that, like the Baronne de Montesquieu, she handled the household accounts.[8] As Madame Ampère was frequently ill, she had to find money to pay for various doctors' services and remedies. Difficult though her situation seems to have been, Julie maintained total control of all domestic arrangements. Despite or because of the effort she exerted in managing her household, Julie Ampère showed little to no interest in the details of her husband's scientific and mathematical work.[9] The Ampère marriage, like that of Montesquieu, adopted a gendered division of labor. André-Marie devoted himself to his career; his wife focused on their household.

Family members supported intellectual households in numerous practical ways.[10] Although such workers have often slipped out of the historical record, they were indispensable in enabling intellectual work. Household labor kept the Republic of Letters running smoothly.

## MAKING AND PROMOTING KNOWLEDGE

Other savants saw more than household managers when they regarded their relatives. Jérôme Lalande, one of the most accomplished astronomers of the Enlightenment, proved especially resourceful in putting his friends and family to work. Born in Bourg-en-Bresse in 1732, Lalande secured election to the Academy of Berlin at the tender age of nineteen. Shortly thereafter, he entered the French Academy of Sciences with unanimous approval in 1753. These early successes blossomed into an illustrious career. His works enjoyed international renown, and his popular classes at the Collège de France, his public lectures, and his active correspondence speak to his stellar reputation.

A lifelong bachelor, Lalande delighted in romantic affairs. He often combined his love for astronomy with his love for romance by collaborating with learned women. In the 1780s and 1790s, he worked with his mistress, Louise-Elisabeth-Félicité Du Pierry. Du Pierry studied natural history and astronomy and became the first woman to teach astronomy in Paris.[11] During another affair, he fathered a daughter, Marie-Jeanne-Amélie Harlay, sometime between the years 1767 and 1770. Lalande collaborated with women even if romance was not an option. Of these, perhaps the most significant is Nicole-Reine Lepaute. Lalande found Lepaute, the wife of a well-known clockmaker, impressive in nearly every way.[12] They collaborated for decades, most notably for the *Éphémérides* and the *Connaissance des Temps*, two astronomical publications that required copious calculations.[13] Likewise, Charlotte Amalie, the Duchess of Saxe-Gotha, worked with Lalande by gathering data and performing mathematical calculations, which he incorporated into his published works. "It is a passion of mine to calculate," she enthused.[14] The two were very fond of each other and the duchess referred to Lalande as her uncle, a common way to indicate long-term devotion.[15]

In the last decades of the eighteenth century, Lalande increasingly collaborated with his biological family. His nephew Michel Lefrançais and his illegitimate daughter Amélie Harlay both came to live with him and worked at his observatory at the École Militaire in Paris. In his correspondence, Lalande most often referred to this young woman as his "niece," a generic term for relative. He presumably did so as a way to conceal her illegitimacy from unfriendly observers.[16] Amélie married Michel in 1788, and by 1801, the family included a number of scientifically named children including Caroline (as in Herschel), Isaac (as in Newton), and Uranie (the muse of astronomy, who had recently been honored by having the new planet Uranus named after her).

These relatives remained useful collaborators for the rest of Lalande's life. Starting in 1788, he trained both Amélie and Michel in astronomy, and their labor proved essential. They worked with Lalande on his catalogue of 50,000 stars, an enormous undertaking that he "would not have dared to begin . . . on [his] own."[17] Redacting stars was an arduous process, with each redaction requiring thirty-six separate calculations. Amélie performed these calculations with great speed and dexterity. Her father noted that she "calculates assiduously and has already furnished [me] with more than 3,000 calculated stars." She did so with "a quite uncommon level of skill and facility that makes her extremely useful to astronomy."[18] Calculating was not, for Lalande, the work of drudges and automatons. By his estimation, it was associated with individuals of intelligence, talent, and tenacity.[19]

The Lefrançaises worked in exchange for financial support and lodging. In an allocation given to his daughter and nephew, Lalande noted "the services which Michel Jean Jerome Lefrancais [my] relative has rendered to astronomy and to [my]self personally," including observations and calculations. Lalande continued, "Jeanne Harlay Lefrancais spouse to the aforementioned Lefrancais calculates assiduously." For their services, Lalande granted them "five thousand *livres* . . . payable as long as they live and work for [me]."[20] In a declaration of property written in 1795, Lalande underscored this notion of a family universally dedicated to the pursuit of astronomy. He wrote, "I am a bachelor, but to fulfill the duties of a citizen I have [supported] for many years the children of my relatives. . . . I published this year 300 pages of charts for navigation which attest to our work."[21]

The Lefrançaises depended on Lalande for housing and financial support, which probably chafed at times. Although Lalande praised the talents of his daughter and his nephew, he could also prove a cantankerous taskmaster. In a letter to Du Pierry in 1788, he lamented that "when I am not here, my observatory [boutique] shuts down."[22] The Lefrançaises probably did not appreciate such an attitude and the oversight it fostered. Although Lalande nearly always wrote about his family in admiring terms, his occasional spurts of displeasure show that he had high expectations for their work and could become difficult if they did not meet those standards.

In working with his family, Lalande kept company with many other astronomers. The seventeenth-century astronomers Elisabeth Hevelius and Maria Kirch collaborated with their husbands. The Cassini family had an astronomical dynasty that spanned the whole of the eighteenth century. Caroline Herschel, a much-admired contemporary of Lalande's, worked alongside her brother William.[23] The late hours and intimate quarters of the observatory made this branch of science a particularly good fit for domestic collaboration.

Astronomers may have been particularly well represented in this area, but savants of all sorts worked with their kin. Marie-Anne Lavoisier collaborated with her husband, the chemist Antoine Lavoisier. Her work—which included observing, illustrating, and translating—led one of their friends to christen her an "épouse philosophique," or a philosophizing wife.[24] This was no mere happenstance: the "épouse philosophique" was made, not born. Marie-Anne Paulze was only thirteen when she married Antoine Lavoisier in 1771. Her father, a prominent member of the General Farm (the company commissioned to collect royal taxes), grew alarmed when his uncle, the powerful controller-general of finance Abbé Terray, expressed interest in arranging a match between his mistress's brother and Marie-Anne. As the brother in question was fifty years old and without any substantial income, Paulze felt unenthused about the match. After summoning the courage to turn down his uncle's offer, he forestalled future interventions by broaching the subject of marriage to his young colleague Antoine Lavoisier.[25] In addition to his post in the General Farm, Lavoisier had recently secured election to the Academy of Sciences. An alliance with the powerful and wealthy Paulze family constituted a coup for the young savant, and Antoine and Marie-Anne married with dizzying speed.

After such a brief courtship, they could hardly have known each other. Given the differences between their ages (Antoine had twenty-eight years to Marie-Anne's thirteen), their initial interactions might have been awkward. They seemed to like each other well enough, however, and discovered common ground in Antoine's scientific research. As his bride was still young, Antoine had considerable leeway in completing her education. Whether the scientific bent of Marie-Anne's program of study was Antoine's idea or hers is unclear, although his age and greater experience would have given him the upper hand.

In particular, Marie-Anne studied chemistry. Many male savants came to consider her well versed in scientific matters, as they sent her letters on topics ranging from designs for a new coffeemaker to experiments with burning diamonds.[26] Her understanding of scientific principles also prepared her to eventually act as a secretary in the laboratory, where she recorded results. Her diligence was necessary as Antoine had a penchant for scribbling notes on whatever piece of paper might be handy, including the backs of envelopes and playing cards. Marie-Anne copied his notes into a more durable and organized register.[27] Together with Antoine's students, the Lavoisiers logged considerable hours in their Arsenal laboratory.[28]

Marie-Anne expanded her training to make concrete contributions to Antoine's research. She studied English and Latin so that she might read foreign texts and translate Lavoisier's work. Her brother tutored her in Latin, which she ruefully acknowledged might be boring for him. Nevertheless, she pleaded that he "make me decline and conjugate, which will give me some pleasure and which will make me worthy of my husband. . . . I must know Latin and I am just beginning to grasp it."[29] More significantly, she studied English, the language of some of the most prominent figures in chemistry. She made herself ready to reach out to potential collaborators and to face off with Lavoisier's competitors.

Marie-Anne Lavoisier's education might strike some readers as constricting. Rather than pursuing an independent research agenda, she fashioned herself according to her husband's needs and desires. Removed though it may have been from modern ideals of women's education, the fact that notions of complementarity and companionate marriage undergirded Madame Lavoisier's education does not mean that her education lacked a serious purpose. She went beyond acquiring enough learning to make her pleasant company: she mastered skills that filled gaps in her husband's repertoire, enabled her to collaborate on his publications, and facilitated her own interests in hosting a salon and debating scientific theories.

To be clear, Marie-Anne Lavoisier was no drudge. The vivacious young woman delighted in witty company and often hosted her husband's fellow academicians and savants from around the world. As the Swedish astronomer Anders-Johann Lexell noted, "M. Lavoisier is a pleasing looking young man, a very clever and painstaking chemist. He has a beautiful wife who is fond of literature and presides over the Academicians when they go to his house for a cup of tea after the Academy meetings."[30] That Gouverneur Morris, a chauvinist who disliked learned women, disapproved of her energetic interventions in scientific conversations further suggests that Marie-Anne Lavoisier relished her role as collaborator and conversant.[31] She seemed to have it all: money, looks, charm, intellectual firepower, and a beautiful home in Paris where she could show it all off.

Madame Lavoisier had a particularly active agenda when it came to defending and promoting her husband's reputation. A public relations genius, she developed elaborate schemes to elevate her husband's theories; she even staged productions dramatizing his discoveries in her home.[32] The two languages she chose to pursue, English and Latin, enabled her efforts.[33] Antoine did not speak or read English well, which was a liability given his interests in the work of British thinkers. For example, when Antoine began a correspondence with Josiah Wedgwood in 1791, requesting information about England's natural clays, Wedgwood obliged by sharing the results of his experiments in a letter written in English; Marie-Anne translated this letter into French for her husband. Antoine responded with a letter in French.[34]

Beyond merely facilitating communications, Marie-Anne Lavoisier's mastery of English proved an important weapon in the "combat between oxygen and phlogiston" that raged between Lavoisier, Joseph Priestley, and their respective allies during the 1770s and into the 1790s.[35] On the one side was Priestley, an English chemist of great repute, who maintained that many chemical processes could be explained by the existence of a substance called phlogiston. Phlogiston was so central to Priestley's theories that when he successfully isolated oxygen he called it "dephlogisticated air." Lavoisier, on the other hand, argued that air did not contain phlogiston (a substance whose existence he denied). He was, however, much intrigued by Priestley's experiments with "dephlogisticated air," which he renamed oxygen.[36]

Phlogiston notwithstanding, the two men also differed in their personalities and scientific methods. Both were committed to educating the broader public, and both represented themselves as moral characters who deserved the public's trust. But they had different ideas about the best way to accomplish this goal. Priestley portrayed himself as an honest experimenter who humbly submitted his views for public feedback. He believed his self-deprecating asides, which he called "marks of candour," helped the public see him as someone who understood his limits. Lavoisier, on the other hand, went for a splashier style. He invested huge sums of money into new equipment and made bold claims for his theory of oxygen and the future of chemistry in general. Some observers found him arrogant.[37] Priestley did not just disagree with Lavoisier: he found his intellectual style repellent and attacked the other savant as egotistic and exclusive.[38] At first, Priestley had the upper hand, for the English scientific community mostly supported their compatriot. French savants gave Lavoisier's ideas a warmer reception, but they took a wait-and-see approach.

Having a savvy wife like Marie-Anne Lavoisier made the road to chemical revolution easier to travel. As the rivalry between Priestley and Lavoisier grew heated, Madame Lavoisier did not hesitate to take up arms in her correspondence, publications, and salon. The tenacity with which she fought is revealed in her correspondence with Horace Benedict de Saussure, a celebrated Genevan savant. In 1788, Saussure discovered that a friend of his had informed Marie-Anne Lavoisier that Saussure opposed the theory of oxygen. Fearful of what she now thought of him and anxious to salvage their friendship, he hastened to clarify his position. He had not lobbied against Lavoisier, he assured her: "It is quite true that I am a bit of a skeptic, that I have not found his theory to be entirely proven, that I have mentioned some doubts, but to say that I have dogmatized against M. Lavoisier's system . . . that I find his claims without any foundation . . . this is absolutely untrue."[39] Saussure's letter went on for some length: he admitted his doubts about oxygen but also asserted his impartiality as well as his personal regard for Antoine.

He fretted that rumors of his intransigence would so outrage Madame Lavoisier that she would break ties with him. "I ask you, Madame," he pleaded, "to no longer give me the odious title of declared enemy to your marvelous successes; I am a very zealous admirer of Monsieur Lavoisier and his savant collaborators."[40] In this letter, the tense rivalry between Lavoisier and Priestley made Saussure fear that his alleged sympathy to Priestley's ideas might damage his relationship with the Lavoisiers. It is also apparent that Marie-Anne played an active role in leading the fight against the defenders of phlogiston, for Saussure addressed the letter to her. She must have been a force to reckon with, for his anxiety at the thought displeasing her was palpable. By this point, the Lavoisiers had been married for seventeen years. Marie-Anne had grown confident in her understanding of scientific matters, committed to the mission she shared with her husband, and able to wield considerable influence over scientific networks.

Madame Lavoisier also proved a useful lieutenant when she translated Richard Kirwan's 1787 *Essay on Phlogiston and the Constitution of Acids*. Kirwan was a Priestley ally, and Lavoisier's translation of his theory probably helped her husband stay abreast of current theories. Then the Lavoisiers went one step further: Marie-Anne published her translation of Kirwan's work and included essays by Antoine and his supporters critiquing Kirwan's ideas and advocating their own theories. Madame Lavoisier made certain to send this volume to select contemporaries, including Jao Jacinthe de Magellan and Marsilio Landriani.[41] Her translation of Kirwan's work and its publication alongside Antoine's critique apparently made a compelling case for the latter's ideas. Saussure—who, as noted above, had been skeptical of Lavoisier's ideas—wrote a letter of congratulations upon reading the translated volume: "You triumph over my doubts, Madame, at least regarding phlogiston, the principal object of the interesting work that you did the honor of sending to me."[42]

Saussure did not just consider this a scientific triumph: he also suggested that it was a moral victory and public relations coup for the Lavoisier camp. Antoine and his colleagues suffused their arguments with "clarity" and "nobility." Kirwan's text, however, was marred by "confusion and rage." Saussure's feelings about Kirwan had declined precipitously: "despite the grace and the precision with which you have translated his book . . . [the translation] is harmful to his reputation in bringing to light the sign [of his] weakness and, often enough, the bad faith of his reasoning."[43] Marie-Anne Lavoisier's translation of Kirwan's *Essay* struck the decisive blow against the theory of phlogiston, at least as far as Saussure was concerned. Her mastery of the English language proved a useful weapon in propelling her husband to victory over his British rivals. In part due to her efforts, Lavoisier's theory began to gain widespread acceptance.

Marie-Anne also directed Antoine's communication with Francophone scholars. She punctuated her correspondence with messages to or from her husband. Moreover, she helped savants stay abreast of current developments and new publications; her husband's close friend Magellan mentioned more than once that he appreciated the news and texts that she had passed along.[44] By managing at least part of Antoine's correspondence in this way, Marie-Anne helped her husband sustain the open lines of communication that characterized ideal scholarly exchange in the Republic of Letters. This permitted him to devote himself more fully to his work. He rose at five in the morning, worked on scientific matters from six to nine, worked for the General Farm and the Gunpowder Commission until seven at night, and returned to the laboratory from seven to ten.

Saturdays he devoted to scientific experiments with his students.[45] This
rigorous schedule left little time for letter writing. His wife's ability to
send and receive messages of a scientific nature helped lighten the load.

The most visible contributions Marie-Anne Lavoisier made, however,
were iconographic. A student of Jacques-Louis David, she put her skill to
good use by illustrating Antoine's *Traité elementaire de chimie*, which
included thirteen engravings based on her drawings.[46] Her illustrations
offered readers a glimpse into Lavoisier's laboratory and showed how Antoine conducted experiments. She also included drawings of his expensive,
custom-designed instruments. Such devices remained controversial in
some circles: Priestley had lambasted Lavoisier for his instruments, claiming that their expensive price tag was a sign of their owner's exclusivity.
Marie-Anne's illustrations were thus more than pretty pictures.[47] They
provided the scientific community with an understanding of Lavoisier's
techniques and experiments, and—in so doing—may have helped protect
him against accusations that he desired to work in secret, outside public
view. Her drawings of Antoine, herself, and his students at work helped
bring his experiments to life; see Marie-Anne's *Expérience sur la respiration de l'homme au repos* (figure 2.1) and *Expérience sur la respiration de
l'homme exécutant un travail* (figure 2.2). The illustrations indicated who
did what and how, and therefore opened up the laboratory and Antoine's

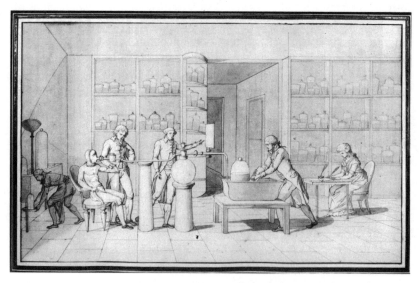

Fig. 2.1. Marie-Anne Paulze Lavoisier, *Expérience sur la respiration
de l'homme au repos.* 1790–1791. Private collection.

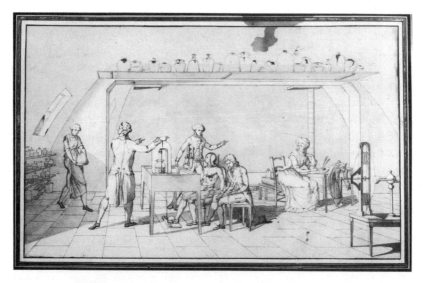

Fig. 2.2. Marie-Anne Paulze Lavoisier, *Expérience sur la respiration de l'homme exécutant un travail.* 1790–1791. Private collection.

experiments to the public. These refined, almost classical compositions infused Lavoisier's experiments with gravitas.[48]

Marie-Anne Lavoisier collaborated with her husband in many ways. Her translations helped him remain abreast of burgeoning theories in England; her drawings helped clarify how his experiments worked; her social network helped promote his ideas. Her work to publicize his theories and defend him as the premier savant of his age proved invaluable at a time when Lavoisier's ideas were hotly debated. She threw herself into this work with much enthusiasm and creativity, and she seems to have derived a great amount of intellectual satisfaction from her work. She was more than a diligent and retiring helpmeet; as a public relations expert *avant la lettre*, she collaborated with Lavoisier and pushed him to advocate his ideas in daring and stylish fashion.

Marie-Anne Lavoisier performed her job so well, in fact, that any struggles or frustrations she may have experienced have been obscured. Perhaps she yearned for more attention and praise than her role as her husband's associate permitted. Perhaps she grew weary of following Antoine's research agenda rather than pursuing her own ambitions. Skilled as she was at representing her marriage as an affectionate and productive match, I can only guess at such troubles. One thing is certain: the Lavoisier marriage was surely more complicated than their reputation suggested. Indeed, Madame

Lavoisier, whose public face was that of devoted wife, indulged in a long-term affair with her husband's friend Samuel-Pierre Du Pont de Nemours.

The story of the Lavoisier marriage and its aftermath raises key points about intellectual families and labor. Monsieur and Madame Lavoisier worked together effectively for years but their marriage and collaboration came to an end in 1794. Lavoisier had been a tax farmer, a profession that haunted him during the Revolution. He had also made a powerful enemy in Jean-Paul Marat, the malicious journalist, when Lavoisier had snuffed out Marat's hopes of publishing his research with the Academy of Sciences. Marat never forgave the slight and kept a careful list of Lavoisier's every misstep. Marat held forth on Lavoisier's many failings in his newspaper, and this did not help the savant's reputation. In 1794, for the crime of having been a tax farmer, Lavoisier was arrested and executed. Marie-Anne was only thirty-five at the time of his death. She herself was arrested, briefly imprisoned, and deprived of her property for some time. Although her marriage enabled her participation in major scientific debates, it also constrained her intellectual career. She had developed skills to complement Antoine's, but their work died with him. Her predicament upon his death underscores the patriarchal nature of family workshops; she had worked by his side for years, but that did not leave her with an easy way to shift from collaborator to independent agent.

Marie-Anne Lavoisier did not stop living her life when Antoine died, however. Her work with her first husband may have been crucial to his science but it did not determine her whole life story. Once the political climate calmed after the Terror, she resumed her salon and hosted an impressive array of guests that included Delambre, Cuvier, and Humboldt. She soon attracted the notice of Count Rumford, who pursued her doggedly for four years. "I can imagine nothing sweeter than to live with you, work with you all day, and then fall asleep in your arms," he wrote.[49] Perhaps hoping that she could rekindle the same sort of collaboration that she had had with her first husband, Lavoisier accepted his proposal in 1804. The match was a disaster, however. Rumford wanted peace and quiet; his new wife wanted to socialize. They could not meet in the middle.[50] Tired of her husband's heavy-handedness, Marie-Anne eventually took action guaranteed to get his attention: she poured boiling water over his beloved rose bushes. They separated shortly thereafter, and Marie-Anne resumed a busy social schedule until her death in 1836. She lived a full life after Antoine's execution, although she lost the combination of scientific collaboration and social prominence that she had enjoyed as Madame Lavoisier.

Antoine reaped more straightforward benefits from their marriage. Ideally, eighteenth-century savants were not supposed to act like self-serving publicity hounds. Savants had to walk a tightrope: balancing the need to adhere to norms of politeness and community, on the one hand, and arguing in favor of one's ideas, on the other. This could be a tricky business, but marrying well made it easier. This was a time when savants stepped into a new role as moral authorities and—in some cases—celebrities. Renowned for their moral probity and admired for their brilliance, famed savants were stars of their era, and their wives were their publicists. Although Lavoisier certainly defended himself, the fact that his wife was so vigilant on his behalf helped Antoine distance himself from what Dorinda Outram has called "dirty power."[51] Lavoisier was still accused by his critics of being arrogant and self-aggrandizing, but these accusations may have had a greater impact if his wife had not shouldered so much responsibility in advocating his ideas.

Marie-Anne Lavoisier was not the only wife gifted in public relations. Suzanne Curchod Necker, wife of the finance minister and enlightened reformer Jacques Necker, proved a valuable ally in her husband's fight to reform the government and win over public opinion. Both Jacques and Suzanne skillfully manipulated public sentiment.[52] They shared an interest in hospital reform, which they believed demanded immediate action. The hospitals of the time, often dirty and overcrowded, were out of sync with Enlightenment ideals of order, sympathy, and privacy. Most facilities had stultifying interiors, patients packed two or more to a bed, and environments that facilitated the spread of contagious diseases.[53] Some Enlightenment thinkers were so appalled by these institutions that they recommended their universal closure, arguing that patients would be safer and more comfortable at home.[54] Necker, in his capacity as finance minister, proposed to transform hospitals by making them more sanitary and efficient. Naturally, new doctors and new facilities would come at a price, but Necker was not one to be dissuaded by a high price tag. He proposed an ambitious financial scheme in which hospital administrators would fund reforms by selling excess land and investing profits in government *rentes*. Yet few administrators followed his suggestions. Jacques had hit a dead end.

His wife, however, had more success with her own efforts to reform the capital's hospitals. Like many women of her station, including a host of marquises, countesses, minister's wives, and the always-active Marie-Anne Lavoisier, Necker devoted much time to hospital reform.[55] She developed a reputation for being dedicated to charitable causes and eventually assumed control of a new hôtel-Dieu in Saint-Sulpice. At Saint-Sulpice, Madame Necker enacted reforms on a microlevel that her husband hoped to implement on a grander scale. She insisted that each bed house no more than one patient. She demanded that all rooms be well ventilated and that patients' medications and diets be carefully regulated by trained doctors and nurses. All of this was in line with Enlightenment ideals of charity and medical improvement. And, ever the efficient administrator, Madame Necker even managed to cut the operating costs of the hospital. The hospital of Saint-Sulpice became a showroom of enlightened medicine and attracted visitors from near and far. The government even published a pamphlet detailing the methods employed at Saint-Sulpice in an explicit attempt to make the institution a model of enlightened hospital reform, and at least one provincial hospital sprang up in imitation of Madame Necker's masterpiece.[56]

Jacques Necker himself advertised the success of the project and his wife's role in bringing it to fruition in his 1781 *Compte Rendu au Roi*. This publication was a smash best seller in which Necker discussed the state of royal finances with an unprecedented level of transparency (although he artfully disguised the kingdom's financial weaknesses). Necker included a section called "Hospitals and Prisons" in which he detailed various reforms enacted in the spheres of medicine and criminal detainment. He applauded the general success of the initiative to improve hospitals and concluded the section by noting that he felt obliged "to indicate, without naming her, a person graced with the rarest of virtues and who has helped me very much to accomplish Your Majesty's agenda." He went on to note that "it is precious for a Minister of Finances to have found in his life's companion a safety net for all the details of good works and charity that escape his attention and abilities. . . . He is thus happy to find that the particular needs of the poor and the miserable may be met by an enlightened person near to him who shares his sentiments and duties."[57] Thus, in his *Compte Rendu*—a publication of great popularity—Jacques acknowledged the important role that his wife had played in the matter of hospital reform and thanked her for her assiduous efforts.

Other publications also noted her work: the *Mémoires secrets* observed that Suzanne "is the general director in this matter; she is at the head of the hospitals, charity houses, and all the other institutions dedicated to alleviate the troubles of humanity. She has in this realm vast projects, like her husband does in his."[58] One month later, this same publication recorded her involvement in the reform of the Conciergerie prison.[59] Her charitable work had such positive connotations that it was difficult for the Neckers' many enemies to find fault with her work (although some gave it a shot).[60]

Suzanne Necker's work on hospital reform complemented her husband's work as finance minister. She did what her husband could not,
and she did so in a way that amplified their public appeal. As Protestant
foreigners whose reform agenda irritated many grandees, the Neckers depended on public support. They needed to generate publicity but without
stirring up concerns that Jacques was arrogant and Suzanne ambitious.[61]
Suzanne Necker surely cared about hospital reform, but she also knew
that her work would reflect well on her. Organized, efficient, and seemingly motivated by pity and sympathy, she struck all the right notes. For
having had the good sense to marry her, Jacques's reputation also profited.

Suzanne Necker's many talents included a knack for conversation, and
she ran a popular salon and correspondence network. Her salon attracted
intellectual luminaries and aristocrats.[62] She used her influence in this
sphere to her husband's benefit. On one occasion, she read aloud an essay of her own composition titled "Portrait of Mr. Necker," whom she described as "you alone [who] are always the object of all my affections!" She
then embarked on a highly flattering description of her husband's intelligence and virtue. Monsieur Necker was "born a thinker," he was "disinterested" and not concerned with personal profit; his only interest was
to serve the public good. Her vantage point gave her an ideal view: "I have
formed a sound understanding of Mr. Necker: having become his happy
companion, it was easy for me to judge his virtues. He has filled my life
with so many different kinds of happiness."[63] That she read this essay in
her salon is significant, because salons were a key venue for Enlightenment sociability and political maneuvering. As Necker drew a posh crowd
to her home, the salon was an excellent venue for generating good press.
Some may have dismissed her praise out of hand—she was married to the
man, after all, and was hardly an objective observer—but others may well
have been moved by her eloquence and affection.

This essay, written in 1787, was Madame Necker's attempt to influence debates about her husband's work as finance minister. Necker had
first been appointed to the post in 1776 and had been dismissed in 1781.
He remained a visible and controversial figure long after his dismissal, in
part because of his publication of the *Compte Rendu au Roi* and *Traité de
l'Administration des Finances en France.* Aided by some sly accounting,
he had painted a rosier picture of French finances than many had been
expecting. These pleasant surprises made Necker into an overnight sensa-
tion, a real public hero. King and court felt much less warmly toward him,
however. Many at Versailles grew irritated by Necker's self-confidence
or concerned about his Protestant faith. Various court factions, most no-

tably that of Queen Marie Antoinette, opposed Necker. Moreover, many philosophes resented Necker, particularly because they remained loyal to his predecessor, Anne-Robert-Jacques Turgot. Even "the good" Condorcet could turn positively vituperative when he wrote of Necker. Besieged on all sides, Necker found resignation to be his only option in 1781. His dismissal did not end his story, however. Jacques Necker continued to defend his ideas, and his opponents continued to lampoon them. Suzanne Necker also came under fire, with critics lambasting her ostensibly single-minded dedication to her husband's career.

It was in this context that Suzanne read her "Portrait" aloud. By describing Jacques as a disinterested servant of the people who felt compelled to speak his mind, she defended him against charges of greed and ambition. She attempted to bolster his already considerable popularity with the general public. Although she did not publish her essay, choosing to read it aloud in the confines of her own home, she did expect her guests—who often included such luminaries as the essayist Antoine Léonard Thomas and the natural philosopher George-Louis Leclerc Buffon—to share her good opinion of Monsieur Necker. She hoped the essay would motivate her friends to defend Necker and his policies beyond the confines of her salon. The "Portrait of Mr. Necker" was public even if it was not published.

Suzanne Necker's projects helped her husband in more ways than one. Her efforts to rebuild the hospital of Saint-Sulpice and make it a site of enlightened and compassionate medicine enabled her to accomplish reforms that her husband had been unable to realize. Not only did her work help Jacques reform one hospital, but her success also afforded him proof that their ideas were sound and allowed him to advertise their success to the king and the public via the *Compte Rendu*. She likewise used her salon as a venue in which to defend her husband's reputation. More generally, it behooved Jacques Necker for his marriage to appear enlightened and productive, just as it did the savants discussed in the previous chapter. It enhanced his reputation if he appeared to have a virtuous and enlightened spouse and for it to seem as if they worked together for the public good. The Neckers thus cultivated a shared public image of virtue and enlightened reform.[64] Jacques displayed his loving and productive marriage as proof of his virtue, utility, and sensitivity—all characteristics that he believed qualified him for public service.

Marriage clearly had its perks for Jacques Necker. Suzanne fulfilled the post of minister's wife with aplomb and used her considerable wit, charm, and administrative talents to advance their reforming agenda and to defend her husband's reputation. But like Marie-Anne Lavoisier, she paid an opportunity cost. She devoted herself to her husband's career, which entailed sacrificing her dreams of authorship. Necker wrote constantly, filling journals and keeping up a voluminous correspondence. Unfortunately, her husband Jacques considered female authors unseemly. As her daughter Germaine de Staël later noted, "I am persuaded that she would have acquired a very great reputation for intelligence; but my father could not stand a *femme auteur* . . . . Mama had a strong taste for composition, she sacrificed it for him."[65] Suzanne Necker published nothing during her own lifetime. She pursued her ambitions and passions in various ways, but the route of published authorship remained closed to her.

Her daughter's life turned out quite differently, and the contrast is telling with regard to the choices available to learned women at the end of the eighteenth century. At twenty, Germaine married the Baron von Holstein, a Swedish nobleman of no particular fortune. Theirs was not a love match. De Staël needed a husband for the sake of propriety, and she appreciated her spouse's diplomatic standing, which afforded her a certain degree of social prominence and political cover. Unlike her mother, Germaine de Staël did not devote herself to her husband's career; she had plans of her own. She published her first book in 1785 and published dozens more during her lifetime. She also pursued an active political agenda. She had a liaison with Benjamin Constant, a like-minded political figure. But in neither this nor any other relationship did Germaine de Staël depend on a man for her public stature. She was the star. The result was, compared to her mother's life, a more autonomous existence as a learned woman but a rockier path of exile and controversy. Today, Germaine de Staël is regarded as a brilliant novelist and a daring political thinker; her mother is much less known.

Germaine de Staël was not the only woman of letters to publish her own work, of course. Even among those women married to men of letters, some managed to pursue their own intellectual agendas. Sophie Grouchy de Condorcet, for example, published her own work. Like many of the wives discussed in this chapter, she worked with her husband and hosted a salon that helped develop his reputation and publicize his ideas, as well as her own. Condorcet died in 1794, however. Condemned for his dissenting views during the Terror, he went into hiding for several months. He was eventually apprehended, and died under mysterious circumstances while in prison. Grouchy de Condorcet survived the Terror by dint of her talent and determination: a gifted artist, she painted portraits to cover expenses. Once the threat of the Terror had passed, she edited Condorcet's unpublished works. But her plans did not stop there. In 1798, for a project more her own, she translated Adam Smith's *Theory of Moral Sentiments* into French and published her own letters on the subject. She may well have pursued this work even if she had still been married to Condorcet—he was a feminist and might have encouraged her—but her status as a widow gave her the freedom to pursue projects at her leisure. Marital collaboration was not the only option available to learned women.

Anne-Catherine de Ligneville Helvétius, wife of Claude-Adrien Helvétius, was herself an active collaborator and, later, an independent widow. The Helvétiuses enjoyed a lofty aristocratic status and they could afford to pursue their intellectual passions. Anne-Catherine ran a popular salon, and Claude-Adrien dabbled in Freemasonry and wrote works of materialist philosophy. In his notorious book *De l'Esprit* (1758), he denied the existence of an immortal soul. Although this work did not correspond with mainstream thought, it initially passed muster with royal censors and received the king's imprint. This meant that a book that challenged Catholic dogma was circulating with the royal stamp of approval. Before long, the *directeur de la Librairie*, Malesherbes, received complaints from Jesuit professors at the Sorbonne that the book trafficked in immoral theories. How had it had been approved for publication? A political firestorm ensued. The Paris Parlement issued a rousing condemnation and ultimately revoked Helvétius's permission to print his work. He attempted to mollify the Parlement by recanting his ideas but became so alarmed by the judges' rancor that he left Paris for his estate of Voré. He worried that the next step for his critics involved throwing him into the Bastille prison.[66]

From 1758 to 1761, Claude-Adrien Helvétius found himself in a dire situation. In this dark time, he trusted his wife above all others. As he wrote to her: "You are my consolation; at least I have one person who loves me. Ah! what happiness to have such an estimable wife during such unhappiness. . . . I cried while reading your letter, and in recalling how much I love you and how much I am loved."[67] A few weeks later, he reiterated these sentiments: "Oh! that I have seen friends turn their backs on me! . . . But you remain with me; it is on you that my happiness relies, and I will never be without you."[68]

Madame Helvétius did more than provide emotional support. While her husband was in exile at their estate in Voré, she defended his work before the Parlement of Paris to protect him from further censure. Her testimony, which she delivered in person sometime between 23 January and 3 February 1759, attested to her husband's good character and morals. She also reminded judges that Claude-Adrien had applied for and received all necessary legal permissions to publish his work. Her husband, a "a zealously virtuous citizen" had founded *De l'Esprit* on "love of humanity and of the fatherland." She pleaded for them to let the work go forward, and railed against the theologians who sought to ban the book. "If one notes that the book has been condemned by bishops, [I] respond by asking: were not the works of Aristotle and Descartes also adopted but then rejected by the Sorbonne? Have not all great men been persecuted by theologians?"[69] Her efforts came to naught, however: Parlement condemned the book shortly after her speech and ordered it burned.

Perhaps recognizing that he needed people other than his wife to vouch for him, Madame Helvétius leaned on her correspondents to defend Claude-Adrien. She wrote a letter to Malesherbes asking him to protect the book. Malesherbes was known to be friendly to Enlightenment writers, inasmuch as was sensible for a royal official, and Helvétius hoped to persuade the director to pull a few strings on her husband's behalf. She assured him that her husband's book was well worth his trouble, as its merits were clear: "The number of editions that have been printed, the letters of praise that my husband receives every day from foreign and French readers alike, people that he has never met, assures his glory." Furthermore, she complimented the *directeur* for his willingness to act in the public's benefit by introducing works of great reason and erudition, and entreated him to protect one such work from the religious zealots of the court. "You are the magistrate of the Librairie," she wrote. "You would like for the public to be enlightened, for their own happiness. Permit, therefore, honest critiques of this work, however strongly worded they might be, but defend us from the insults which might reinvigorate attacks from the devout faction."[70]

Determined to galvanize support, she praised Claude-Adrien's work in her correspondence. In a letter to the historian Jean Lévesque de Burigny, Madame Helvétius thanked him for the "marks of friendship" he had shown in defending *De l'Esprit* and its author. She added: "It is strange that, at the same time that my husband receives so many marks of most-flattering esteem from London, he is exposed to mockery in Paris."[71] Shortly thereafter, she urged Burigny and all men of letters to stand against such an egregious attack on intellectual liberty: "As long as bigots arm themselves against men of letters, the *gens de lettres* must support each other, especially those who, like you and my husband, are a credit to their nation. In order to give you ammunition for this fight, I want to send you a copy of reviews of *De l'Esprit* written in London and Germany."[72] To this letter, Helvétius attached positive reviews of the work.

In these letters, Madame Helvétius cast the persecution of *De l'Esprit* as an attack on reason by religious zealots. According to her logic, Monsieur Helvétius had suffered insults from the devout party because they feared his genius and the changes he might effect in public opinion. Her read of the situation fit with the new model of the philosophe popularized by Voltaire, who tended to portray himself as honest and virtuous and his critics as enemies of liberty and reason.[73] Helvétius framed her husband's struggle as one between reason and ignorance, progress and regress. The fact that the letters came from her, rather than her husband, permitted Monsieur Helvétius to rest above the fray while his wife handled the business of circulating news of his successes and stirring up outrage over the injustices he suffered.

In defending her husband's reputation, Helvétius acted in ways similar to Mesdames Lavoisier and Necker. And as a woman of high aristocratic rank and financial comfort, she enjoyed great liberty and security in the years of her widowhood. Upon her husband's death in 1771, she continued to run her salon and helped found the lodge of Neuf Soeurs. She seemed content with life and was in no particular hurry to remarry. She had money and time on her side.

That Lavoisier, Necker, and Helvétius would protect and promote their husbands' ideas fit within existing models of patronage. Learned and aristocratic women often supported men of letters they admired, and they had done so for centuries. Adept at both verbal and written communication, schooled in manners and charm, and well connected through the salon network to a range of powerful and brilliant individuals, the urbane women of Paris exerted considerable cultural influence. Yet the learned women discussed here protected their own husbands, rather than protégés of another sort. Female family members even protected a savant's reputation after he died, publishing memoirs, correspondence, and completed works to secure a thinker's glory for posterity.[74]

As these examples show, family labor helped make, promote, and defend knowledge. Wives and children calculated, observed, and proselytized. Such work was not restricted to scientific families: philosophes also benefited from the efforts of their wives and children to advertise and defend their work. Women family members helped their male relatives in many ways but proved especially useful with regard to advertising and defending the finished product. Men clearly profited from this arrangement and their family members enjoyed some benefits, including praise for their talents. But there were drawbacks for collaborators. Wives and daughters generally did not embark on autonomous careers. Most wives, for better or for worse, were bound to the family.

## REPRESENTING WOMEN AS LOVING AND LEARNED

Autonomous or not, family members did not necessarily labor in obscurity as invisible assistants. Sentimental savants boasted that their wives and children excelled in every way: they possessed talent, worked diligently, and loved their families. Men of letters depicted themselves as productive and consoling, rational and emotional, and they did the same for their wives. Women, like their husbands, reveled in the warmth of family love and developed cutting-edge ideas. The self-fashioning strategies developed by men of letters—emphasizing their sentiment, virtue, and commitment to public utility—were easily adapted to represent their family members. It was not enough for men of letters to describe themselves as the ideal combination of sentiment and reason; they also wanted to depict their families as loving and learned. They represented their family lives in a way that reflected well on them: the sentimental savant surrounded by his equally sentimental family.

That men of letters would portray their families in affective terms makes sense during the sentimental eighteenth century. A sensitive, warm person with a deep connection to his or her family was a person with good morals. Eager to claim this title for themselves and their loved ones, eighteenth-century individuals drew on emotive, exaggerated language when describing their home lives. Outside the written page, fashionable individuals displayed their sensitivity with copious tears and dramatic gestures.[75] In less dramatic fashion, men of letters often wrote of their connection to their collaborators in affective terms, stressing the "bonds of pleasure and sentiment" that united them.[76] Within and without familial contexts, therefore, eighteenth-century individuals were keen to emphasize their sensibility and their emotive ties to others.

In part, using sentimental rhetoric to describe learned women helped guard them from ridicule, a danger made apparent by the experiences of previous *femmes savantes*. The late seventeenth and early eighteenth centuries had witnessed an extraordinary cohort of women in science. Intrigued by the fashionable new sciences, well versed in foreign languages, and emboldened by feminist assertions that the mind had no sex, an impressive number of women stepped forward to claim a place in the Republic of Letters.[77] Celebrated women like Émilie Du Châtelet (1706–1749) and Laura Bassi (1711–1778) showed how high *femmes savantes* could rise. They assumed public positions as women of science—Du Châtelet through her extensive correspondence networks and publication record,

Bassi through her lectureship at the University of Bologna. Both women mastered Newtonian physics and, in Du Châtelet's case, supplemented Newton's theories with Leibnizian metaphysics. They leveraged their command of the material into careers.[78] Yet Du Châtelet and Bassi pushed the boundaries of what respectable women were supposed to do, and they sometimes felt a backlash. Du Châtelet was accused of stealing ideas from men, lampooned for her frivolity, and ignored as an outsider unable to gain access to the Paris Academy of Sciences.[79] Bassi lived under intense scrutiny, attracting both adulation and accusations of sexual impropriety.[80] Hoping to avoid similar attacks, many women chose to work in full or partial anonymity. Mariangela Ardinghelli, for example, worked under a veil of anonymity that she lifted only under certain circumstances.[81]

Writings by Enlightenment men on the topic of learned women show why women might have hesitated to go public with their ideas. Jean-Jacques Rousseau's *Emile* (1762) demanded that women focus their energies on domestic concerns. Women's only purpose, he insisted, was to please their husbands and nurture their children.[82] In his 1772 *Encyclopédie* article "Femme (morale)," Joseph Desmahis characterized women as inherently emotional beings, as the counterbalance to man's rationality.[83] These "natural" inclinations made women well suited to family life, as they would govern their homes with sweetness and self-abnegation. Even Boudier de Villemert, who criticized those who proscribed women from learning, recommended that women be educated just enough to entertain their husbands and be useful to their households.[84] Being moderately well read and pursuing domestic tasks like weaving would make French women more virtuous, he claimed. Although Desmahis and Boudier de Villemert recognized the contributions of a few women of science such as Du Châtelet, they described these women as exceptions to the general rules guiding women's conduct.[85]

Enlightenment Europe thus held many contradictions for women eager to pursue scientific studies. Seemingly more women than ever before were acquiring scientific educations and were publishing their ideas, but women of science faced considerable risk of ridicule for their intellectual ambitions. They might be mocked as silly creatures or dismissed as cold and unfeeling. Moreover, these texts generally defined gender roles as natural and therefore immutable.[86] Nature, the ultimate guide for Enlightenment writers and readers, seemingly intended women to devote themselves wholly to their families, to be subsumed by love for their husbands and children (or so these male authors claimed). For some writers, like Rousseau, a total devotion to family precluded abstract or rigorous study.

If a woman attempted to become learned or to publish her work, she was necessarily distracted from her true purpose as a wife and mother.[87]

Yet the binary of domestic and intellectual work was not stable. Before and after Rousseau wrote *Emile*, women savants proved adept at using domestic rhetoric to justify their studies and publications. By representing themselves as loving wives and doting mothers, women of science could protect themselves from at least some of the insults that might be lobbed their way. Unfeeling? Clearly not, as they loved and were loved by friends and family. Silly? Nonsense, as their work was useful to their family. Monstrously ambitious? Wrong again, as they had only published at the behest of those close to them. Some men may have had their doubts about learned women, but women proved resourceful at fashioning themselves in ways that facilitated their entry into academic debates and guarded them from social ignomimy.[88] Among the most famous of these women was Caroline Herschel, who claimed credit for her discoveries but also insisted her work was secondary to the accomplishments of her brother, whom she served as a devoted assistant. The emphasis on modesty and domesticity only intensified in the years after the French Revolution. Many were eager to correct for the excesses of the Revolution and, *inter alia*, railed against the outspokenness and relative liberty of women. Rather than speaking out about politics or spending their time at balls, political commentators urged women to spend nearly all their time at home caring for their husband and children.[89]

These contexts shaped the manner in which savants recognized and celebrated their female relatives' work. Loathe to expose their family members to criticism, they depicted women in a way that fit within the constraints of feminine modesty and domesticity. Observers praised Marie-Anne Lavoisier as a learned and devoted spouse. The extraordinary portrait the couple commissioned from Jacques-Louis David in 1787 fed such ideas. David's composition cast a warm glow on the household, where love and learning seemed to coexist happily. Helvétius effused that his wife's efforts on his behalf demonstrated her love and loyalty; Suzanne Necker cultivated a reputation as a loving and helpful companion. These representations spoke to one model of the *femme de lettres*: tireless, talented, and devoted to her family.

A telling iteration of this ideal appeared in Louis-Pierre Manuel's *L'année françoise* (1789). In an article on Philibert Gueneau de Montbeillard (a natural historian), Manuel wrote that Montbeillard had left behind "a wife as commendable through her virtues as she was precious to him through her work. Learned in several languages, trained in various fields, she spared her husband much research . . . and he always wanted her to help him express his thoughts." Their intense affective attachment fueled their intellectual collaboration: "They lived together in the most equal intimacy. . . . They had the same tastes, the same passions. Always in agreement, if one said *no*, it was only so that they might occasionally be two [instead of one]." Manuel held up Madame de Montbeillard as a loving partner in every way: she not only cared for her husband but also collaborated with him. Her translations, writing, and research were indispensable. They lived and worked together so seamlessly that the two seemed to have melded into one. Also admirable was Madame de Montbeillard's (allegedly) singular focus on her husband's happiness. She needed no attention for her own talents, and her modesty led her to downplay her own role, "of which she never spoke. Her only vanity was her husband's happiness."[90] Love, devotion, and labor: these were the key attributes of the ideal collaborator-wife. This portrait and representations of learned wives more generally were not about women as they really were. Instead, they were idealized portraits of the perfect wife.

This formula was common in academic eulogies and biographies. Georges Cuvier described the naturalist Guyton de Morveau as assisted in his translations by "a cherished Spouse."[91] Condorcet painted a warm scene of domestic productivity when he wrote of the Swiss anatomist and naturalist Albrecht von Haller in his library, surrounded by friends, students, children, and his wife, "in whom he had inspired a taste for the sciences."[92] Manuel wrote that the Marquis de l'Hopital "was mourned by a wife who, through her respect for and attachment to him, had become almost as learned as he was. The desire to please him gave her his tastes and a bit of his genius."[93] The loving wife of a savant seemingly did more than tend to his hearth and home; she also collaborated with him in whatever way he required. Wifely duties and affection did not necessarily clash with intellectual work. Women who worked with their families were idealized as loving and dutiful wives and mothers who possessed acumen and skill.

These representations were not disinterested portraits or straightforward reflections of reality. Instead, they added another facet to male savants' self-fashioning. The virtues of a household shone back on its head. For a savant to be the patriarch of a loving, learned family reflected well on him. The patriarch remained the central figure in this social vision: he trained his wife and children; he organized the family's research; he interpreted their findings. This ideal, in turn, involved an important revision of the learned woman. Such representations implicitly suggested that women could and should pursue intellectual work, but only within a patriarchal framework.

The example of Jérôme Lalande makes a particularly rich case study for this idealized vision of female savants. He published extensively on the accomplishments and talents of the women in his life. His female associates—his friend Nicole-Reine Lepaute, his daughter Amélie Lefrançais, his mistress Madame Du Pierrry—made regular appearances in his *Bibliography of Astronomy, with a History of Astronomy from 1781 to 1802*, and he highlighted their work in his *Astronomie des Dames* (first published in 1785, reprinted in 1795 and 1806).[94] He expressed open admiration for his female associates as collaborators, wives, and daughters, although his attributions could be uneven and incomplete. He did not credit his daughter in all his works, nor did he list her as a coauthor or groom her as a protégé, as he did her husband.[95] His example thus makes clear the complicated, sometimes contradictory ways in which savants discussed women's work.

Nicole-Reine Lepaute merited especially lengthy entries for her work. Lalande commended her assistance with calculating the positions of stars and comets and found her an indispensable collaborator on their *Ephémérides* of 1774.[96] Throughout the *Bibliography*, Lalande heaped praise on Lepaute's intelligence, even going so far as to say that she "was the only woman in France to have acquired a true understanding of astronomy," and that she was "an object for emulation for a sex that we are interested in associating with our work."[97] She worked with him for many years.

Lepaute was married to a well-known clockmaker, and Lalande wrote admiringly of how excellent a wife she was; he did not reserve such praise for his own household. Although Lepaute was a prolific calculator, "her calculations did not prevent her from seeing to her household affairs; account books rested alongside astronomy tables." Domestic and intellectual roles commingled to mutual benefit; "how the qualities of the heart add to the glory of the talents of a sharp mind!"[98] Lepaute seamlessly integrated her responsibilities as a wife with her work as an astronomer; there was space enough on her desk for both. Her love for her husband and her passion for astronomy sustained one another.

This representation of Lepaute matched other sentimentalized depictions of *femmes savantes* as loving and learned. Lalande followed a similar path when writing about his daughter, Amélie Lefrançais.[99] He mentioned her facility with astronomical calculations at several points in his text and characterized her work as difficult and time-consuming.[100] Lefrançais had proven herself both dedicated and talented, he reported, and she "zealously assisted her husband with observing and calculating. Two or three hundred of the stars [in the star catalogue] were the product of one very cold and painful night." Very little could slow her down, it would seem. She continued to pursue her "immense work, to which she devoted herself with courage, and which even her pregnancy did not interrupt."[101] Lefrançais's femininity was no barrier whatsoever to her work. In expressing this view, Lalande parted ways with sexist savants who represented women as weak, passive, and constrained within a fragile feminine form. In Lalande's telling, his daughter's tenacity and talent enabled her to overcome the physical and intellectual challenges posed by astronomical work. Even pregnancy—a condition so often highlighted as proof that nature did not intend women to pursue advanced studies—did not stop her. Furthermore, by highlighting her pregnancy, Lalande made clear that astronomy need not interfere with a woman's domestic duties: she could do both.

Lalande did not consider Lefrançais or any of his female associates to be mindless machines.[102] Although his daughter was the observatory's main calculator, he never described her work as simplistic. Instead, he nearly always praised her with words like "assiduous," "zeal," and "courage." "Zeal" and "courage" hardly evoke a mechanistic understanding of LeFrançais's skill. Moreover, he had originally contracted an esteemed male mathematician, Chompré, to calculate the tables for the *Abrégé de Navigation*. When Chompré had to back out, Lalande entrusted his daughter with the job. Chompré's initial involvement suggests that Lalande did not see calculation as essentially feminine or simple.

Such a worthy woman deserved an exemplary spouse, and Lalande lavished praise on the Lefrançais marriage as an ideal combination of love and learning. Amélie's love for her husband, combined with her love of astronomy, gave her the stamina necessary to endure the punishing work described above. He portrayed Lefrançais as her husband's partner, working with "the same zeal" and encouraging him by her example.[103] Their marriage was founded on love, dedication, and talent. Astronomy had pride of place in their union: Amélie had been "consecrated to astronomy through her marriage." She "then wished to likewise consecrate her daughter from the moment of her birth." Lalande did not regard the birth of this child as a private family affair but instead infused it with great astronomical significance. "This child of astronomy was born on 20 January [1790]," he wrote, "the day when we saw in Paris, for the first time, the comet that Miss Caroline Herschel had recently discovered; we thus named the child Caroline. Her godfather was Citizen Delambre, one of France's premier astronomers."[104] He perceived a reciprocal relationship between the birth of new generations and the generation of new knowledge: Lefrançais's love for her daughter and love for astronomy shaped her decision to "consecrate" not only herself but also her child to astronomy, presumably to the science's benefit as well as to her own. That these descriptions of the Lefrançais family appeared in an astronomical bibliography, not personal letters, underscores Lalande's desire to flaunt his relatives, including his daughter, before a broader public.

Lalande also shared such musings in his private correspondence. In a poem entitled "For my daughter," included in a letter to Lefrançais, he wrote that she could become very "absorbed" in her work, and that one might therefore worry that she would "forget her husband, her uncle, her children." In the next stanza, however, he asserted that such fears were unnecessary, for "nothing distracts her from these tender sentiments / She is loved, she is always loving." The poem then celebrated the happiness Amélie experienced while scanning the night sky. In a letter written in October 1793, he again described her along these same lines: Amélie was blessed with "wit, grace, and talents. There is nothing left for me to say, except that she loves her children."[105]

These affective ties further blurred the lines between public and private. In sentimentalized depictions of women astronomers, a woman's love for her family and her love of science coexisted in harmony. Women's alleged emotionality and their domestic duties did not compromise their scientific work but rather facilitated them. Male savants need not assume that women were unfit for astronomy because of any sort of mental shortcoming, physical weakness, or emotionality. Yet these representations did not advance an egalitarian vision of astronomy but instead articulated a complementary and gendered division of labor. Women observed and calculated, but men continued to direct them. Even a celebrated discoverer of comets like Caroline Herschel struggled to break free of this hierarchy. No matter how famous she became, her title forever remained that of assistant to her brother.

In the *Bibliography*, Lalande featured women's love for their families to a much greater extent than he did with male astronomers. As Sarah Ross has noted, men and women used familial language to make the unusual more palatable.[106] Sentimental rhetoric was a protective strategy as well as a reflection of Lalande's views on family love and intellectual work. Lalande made clear that although these women were learned, they were also well-beloved as wives and mothers.

Although this rhetoric may have spared female associates from some criticism and generated praise for their intelligence, some women still preferred the safe cover of anonymity. Ambition was considered a dangerous quality for women and fears of being accused of overreaching led many women to downplay or to hide their intellectual work.[107] Women who appeared eager to collaborate with—or, worse, criticize—learned men courted ridicule. Feminine popularity, visibility, and independence rubbed many men the wrong way, and they were quick to respond with sexual slander and mockery. Anonymity, whether complete or partial, safeguarded against this unpleasant outcome. This was a common stance for women, even when their male collaborators were willing and eager to credit them.[108]

For example, Louise-Charlotte, Duchess of Saxe-Gotha, worried about her reputation and what would happen if word leaked out that the duchess fancied herself a *femme savante*, a fear fueled by her aristocratic status. As such, she pleaded with Lalande to keep her contributions anonymous. Her anonymity was not the result of Lalande's ambivalence, as he noted in the *Astronomie des Dames* that she "ha[d] done a number of calculations but she d[id] not wish to be cited."[109] Saxe-Gotha's aristocratic status made her especially vulnerable to criticism, and she remained vigilant about policing her own modesty. She was not the only female assistant to insist on anonymity.[110] Modesty was an attribute of the ideal woman; love, talent, and hard work were also valued qualities. The need to always be modest underscores how difficult it was for women to claim a status equal to learned men; it was challenging to simultaneously downplay and highlight their contributions.

As these examples show, emerging ideas about sentimental science led to the development of a new emotional framework for intellectual work. As they did when representing male sentimental savants, men portrayed affect and intellect, public and private in dynamic relationship when describing their female associates. Women's love for their families and family-like friends easily shaded into a passion for science. The intimate nature of their work possessed emotional and social meanings beyond its contributions to philosophical and scientific knowledge. This fusion represented an innovative revision of the ideal scientist, the ideal woman, and ideal scientific practice. To a certain extent, such representations of learned women countered stereotypes that women were unfit for advanced study. The family workshop opened up some opportunities for women to work in philosophy and the sciences and to earn acclaim for their work. Yet this sentimentalized vision of women—of *femmes savantes* as diligent and devoted assistants marshaling their many talents to help their husbands/fathers/brothers—was fundamentally patriarchal and therefore put certain constraints on women. They worked in relation to men, not as autonomous figures, and it was accordingly more difficult for them to lay claim to discoveries and innovations.

## CONCLUSION

The story of eighteenth-century intellectual families is, from one perspective, a story of continuity. Families had labored as collective units for centuries, and intellectual families long predated the Enlightenment. In many respects, eighteenth-century intellectual families participated in a long-standing tradition by enlisting their wives and children to translate, observe, calculate, and publicize.

At the same time, Enlightenment families put a new twist on their activities: they represented their work as an extension of their love for each other. This was particularly the case where women were concerned: savants represented women such as Marie-Anne Paulze Lavoisier, Amélie Harlay Lefrançais, Anne-Catherine de Ligneville Helvétius, and Suzanne Curchod Necker as expressing their love for their families through intellectual collaborations. Their love for their husbands and fathers did not conflict with their work; instead, it motivated and sustained them. The family workshop did not foster many independent careers for women, but it crowded women into the threshold between home and academic institutions. Recognized and praised by their contemporaries, the presence of these female family members provided eighteenth- and early nineteenth-century men of letters with many examples of women who were both loving and learned, who conformed to gender norms while still pursuing advanced study.

Family members made a significant impact on public knowledge. They prepared data for public consumption; they facilitated correspondence in the transnational Republic of Letters; they safeguarded a savant's reputation. Although the work performed by family members often took place in private spaces, many savants highlighted such labor in public. They praised familial labor, as Lalande did, or they published illustrations that made clear the collaborative nature of their work, as Lavoisier did. The lines between public and private, affect and intellect, and familial and professional obligations were blurry indeed.

# CHAPTER THREE

# Love, Proof, and Smallpox Inoculation

The cloud of smallpox hung over Europe for centuries, but in the early eighteenth century a ray of hope shone through: Europeans learned from their Ottoman neighbors that they could inoculate themselves against the disease. Strangely enough, the technique seemed to protect people's health by making them sick; it enabled doctors to control the severity and timing of smallpox and to increase their patients' chance of survival. Although many in France had their doubts, inoculation enthralled a sizable minority, especially among the lettered elite. Here stood a clear example of human reason triumphing over the vagaries of illness! Even enthusiasts had to admit, however, that inoculation posed risks. It involved purposeful infection with live smallpox; death, disfigurement, and contagion remained real possibilities. Decades after its introduction in France, inoculation continued to stir up controversy. Although philosophes were not the only group to support inoculation, they were among the most determined. They interpreted the procedure as a harbinger of an enlightened, healthy future, one in which human beings would be spared the random cruelty of disease. Their opponents, on the other hand, lambasted smallpox inoculation as immoral, unsafe, and unproven. Stymied by such criticisms, pro-inoculation philosophes faced an uphill battle.[1]

This chapter is about that battle, and particularly the social visions tied up in it. French philosophers attempted to sway public opinion by telling dramatic and emotional stories about virtuous parents whose "enlightened love" had led them to inoculate their children. They modeled themselves and their family members as "living proof" that inoculation worked, flaunting the virtuous family feelings that had allegedly motivated their decision. Inoculation became yet another way to discuss family life, parental love, and social reform.

Parents had the authority to decide for or against inoculation, and that control shaped the contours of the inoculation debate. Treatises addressed mothers and fathers directly, with writers delving into ruminations on parenthood. Mothers often acted as the driving force for inoculation in their families, and authors sought their support.[2] However, pro-inoculation writers believed that even an enlightened woman did not have full control over her children; eighteenth-century France was nothing if not patriarchal. If inoculation was going to succeed, fathers also needed to support the procedure. The ideal patriarch, from the perspective of the inoculationists, was a loving and rational man who sensibly decided to inoculate his children. His decision was fueled by both reason and sentiment: he opted to inoculate because he loved his children very much and because he perceived that it was the wisest course of action.

For some savants, these were not abstract issues, for they had children of their own. Their parenthood opened up empirical possibilities: they acted as they claimed all enlightened parents should and inoculated their children. Thrilled by the results, they then broadcast their success in their correspondence and publications. They did not treat inoculation as a private matter but rather flaunted it before the public.

Philosophes thus turned ostensibly private medical decisions into public demonstrations. By drawing attention to their families, philosophes infused their texts with the language of feeling *and* generated empirical evidence in favor of the technique. Just as importantly, philosophes styled themselves as models of "enlightened love." They claimed to be ideal fathers: enlightened parents who relied on both reason and emotion in making family decisions. By inoculating their children, savants featured their home lives as laboratories and themselves as good fathers. In doing so, they legitimated their claims about the medical technique and about themselves. By making examples of their children, savants turned the methods of scientific inquiry that they had previously directed toward the study of nature onto their own families. They spotlighted their home lives to show that the right parent—an enlightened parent—could shape the family and society in the way that he or she desired. Family life imbued their experiments with the warmth of sentimental attachment. It allowed savants to perform more roles than one: they could portray themselves as skilled observers and as enlightened father-practitioners.

These inoculations serve as stark reminders that eighteenth-century understandings of scientific inquiry and family love are far removed from our own context. The thought of experimenting on one's own children is shocking to many contemporary readers. Although philosophes saw their

actions as virtuous, their flamboyant inoculations of their progeny show
that they were willing to take serious risks to further their ambitions.
They may well have believed they acted in their children's best interests,
but their reputations were on the line as well. These inoculations are an
extraordinary sign of how far philosophes were willing to go in using their
families to practice their ideas and advance their public standing and intellectual credibility.

The inoculation debates thus shed light on how philosophes sought
to expand their influence over personal family decisions. Pro-inoculation
writers did more than urge the French to inoculate their children; in the
broadest sense, they aspired to create a more enlightened society and to
establish *gens de lettres* and their families as key models for their social
visions. At the same time, sentimental language and the rhetoric of domestic experience legitimated savants' intellectual authority. Philosophes
portrayed themselves as loving fathers who had their family's best interests at heart *and* as selfless patriots who were willing to put their own
children at risk for the greater good.

## THE PROMISE AND PERIL OF INOCULATION

With steady and excruciating progress, smallpox claimed thousands of
lives yearly. The afflicted person first developed a fever. Sores in the
mouth, throat, and nose soon spread over the whole body. These raised
spots eventually formed scabs that made any movement painful. Those
who died generally did so after two agonizing weeks. If they survived,
their fever subsided and their scabs slowly turned to scars. Survivors could
find themselves blind, deaf, mute, or horribly scarred. Worse, victims seldom suffered alone. The afflicted remained contagious for weeks and, as
such, smallpox spread quickly.

This deadly disease alarmed many, but especially those convinced
that France was suffering from a depopulation crisis. Fears of a nation in
decline were actually unfounded: thanks to a reduced mortality rate, the
French population grew over the course of the eighteenth century. But observers were correct in noting a reduced birthrate and smaller family sizes,
and they were alarmed by this development.[3] Casting about for something
to blame, commentators latched onto celibacy, physical and moral decline,
and urbanization as likely causes. Few considered smallpox the driving
cause of depopulation, but the disease's penchant for attacking children
was especially worrisome in light of these fears.

Anxious Europeans spied a glimmer of hope when they learned about inoculation, a process that had been used for centuries elsewhere in the world. Astonished, they discovered that individuals in Turkey, China, and Africa had found a way to shield themselves and their families from smallpox. Onesimus, the African slave of Massachusetts Puritan Cotton Mather, told his master about the technique; around the same time, Lady Wortley Montagu, wife of a British statesman, heard of the procedure during a diplomatic visit to Turkey. Montagu wrote to a friend that the procedure rendered smallpox "entirely harmless" with few side effects: "They keep to their beds two days, very seldom three. They have very rarely above twenty or thirty in their faces, which never mark; and in eight days' time they are as well as before their illness." Compared with the horrors of natural smallpox, the so-called artificial version seemed like a light burden to bear. Impressed, Montagu went on: "You may believe I am well satisfied of the safety of this experiment, since I intend to try it on my dear little son."[4] Women could especially benefit from the procedure, contemporaries argued, for inoculation would leave their beauty intact.

Inoculation is distinct from vaccination, which was not developed until *circa* 1798. Whereas vaccination involved infecting a patient with the harmless virus of cowpox, inoculation was much riskier because a patient was infected with actual smallpox. The procedure most commonly used in France entailed making an incision in two limbs and then inserting a string that contained live smallpox matter extracted from another patient's sores into those wounds. The patient would then contract smallpox, but a much milder version than was usually the case. They exhibited fewer spots, fewer scars, fewer debilitating effects, and a reduced mortality rate. Even before fatality rates could be quantified in reliable mortality indices, individuals had a sense of the major impact that inoculation could have, even if it was not clear *why* the procedure resulted in a milder illness.[5] The operation proffered further advantages in that patients could exercise control over when and where they would acquire the disease. Doctors could ensure isolation and control contagion. They could also verify that the patient was in good health at the time of the procedure, making them better suited to dealing with the disease.

Despite these advantages, however, the decision to inoculate remained difficult. Although the fatality rate was much lower than was the case with natural smallpox, deaths still occurred. There was also the chance that inoculation could spark an outbreak of natural smallpox because inoculated patients remained contagious until their pocks had cleared. Complicating matters further was the fact that physicians disagreed about the value of the procedure. The field of medicine was in flux in the eighteenth

century, which encouraged a host of reactions to inoculation. Before midcentury, the predominant philosophy among physicians was iatromechanism, the idea that the body was akin to a machine. Those adhering to a
mechanistic view of health saw no need for inoculation; in fact, they worried it might be dangerous. After midcentury, however, the mechanists
lost their hold on the medical world, which created an opening for inoculators. Around the same time, medical doctors grew more empirical in their
research and recommendations, preferring to draw on experience and statistics rather than theoretical principles. This development favored inoculation, as there was an increasing body of statistical evidence supporting
the procedure's efficacy. Progress was slow, however, and some physicians
remained suspicious of the technique.[6]

Outside the medical community, the procedure excited and worried
those who contemplated it. Some congratulated themselves for having
the good sense to take a small risk with a big payoff. Such was the case
with Louise Dorothea von Meiningen, Duchess of Saxe-Gotha, who wrote
to Voltaire in 1759. "You will see," she penned, "that, thanks to God, I
have been delivered of my fears about smallpox. My eldest children have
happily survived this cruel malady and the youngest is in the midst of
inoculation. You see that we are very much *à la mode* and free of prejudice."[7] Fashionable or not, others viewed the procedure with more trepi-
dation. Also in 1759, Marie Louise Denis (Voltaire's niece/lover) wrote to
Pierre Robert le Cornier de Cideville that a mutual acquaintance had "an
extraordinary secret which . . . [our friend] has kept it hidden from her
mother, her husband, and from everyone else who cares about her." The
secret? That she would be inoculated the next day.[8] Denis's friend was enthusiastic about inoculation but was surrounded by people who were not;
she had to keep her plans secret lest they try to stop her. Marie Madeleine
de Brémond d'Ars, marquise de Verdelin, also faced significant opposition.
She wrote to Jean-Jacques Rousseau from "my daughters' bedside. I have
finally been permitted to inoculate them." She now felt "the most intense
pleasure in seeing them receive this treatment."[9] Her phrasing—that she
had *finally* been allowed to inoculate—suggests that she had to exert considerable pressure on her spouse before he agreed.

These women all waxed enthusiastic about inoculation, even if they
expressed some concerns. Their letters testify to how vital women were in
promoting inoculation, even if printed texts tended to downplay their involvement. Their support for inoculation can also be explained, at least in
part, by the fact that all three women corresponded with learned men who
tended to approve of the procedure. Confident that they were confiding in

men who advocated inoculation, these women were more likely to frame their experiences as positive and to portray themselves as enlightened for practicing inoculation. These women also remind us that the average beneficiary of inoculation was likely to be of middling or aristocratic social status and a member of the reading public. Inoculation appealed to urban, literate populations much more than it did to rural ones; it was an expensive procedure.

Much of the French population was more reserved.[10] Although the public first learned of inoculation *circa* 1715, it was only in the middle decades of the century that debate took off.[11] Doctors, philosophes, and other interested parties sparred in newspapers, academies, and journals. Interested parties disagreed about the science of inoculation: Did the procedure confer true and lasting immunity to smallpox, or did the effects wear off? Were the inoculated contagious and therefore a threat to public health? Would inoculation cause death or scarring? Still others expressed concern about the ethics of inoculating children. Inoculation presented the greatest benefits to children, many of whom had not yet been infected but probably would be at some point. This posed a moral dilemma, and parents often balked at the idea of infecting their children with a dangerous disease. If the child were to die as a result of inoculation, was it not wrong for the parent to have exposed him or her to the virus? For more reasons than one, many French parents opted not to inoculate their children even as the procedure became common in other European countries.

In this debate, the royal family occupied center stage. Smallpox aroused anxieties about the royal succession, as the French had seen the disease devastate Louis XIV's line. Louis XV, who had emerged the sole survivor of a once-robust group of heirs, feared for his and his children's health. But neither he nor his subjects felt certain about inoculation. To inoculate was to put the lives of the heirs at risk; to not inoculate was to leave them in danger of contracting natural smallpox. The potential consequences of inoculation remained unclear, and Louis XV never did inoculate his children.

Not all princes of the blood felt the same way. In 1756, the Duc d'Orléans, patriarch of the ever-troublesome Orléans branch of the royal family, inoculated his children. That he had to secure Louis XV's permission underscores the political as well as the personal risk he was taking. The inoculations went smoothly, however, and the kingdom rejoiced. The Orléans episode unleashed a wave of popular interest and support. The news sparked celebrations in the street and even spawned commemorative fashions. Supporters cheered Orléans, sported *bonnets à l'inoculation*, wore dresses in the *Tronchine* style (named after Théodore Tronchin, who had performed the Orléans inoculations), and penned poems. "Three times the tenderest of fathers / gave the beneficial order," crowed one author.[12] But, in the end, even this event failed to turn the tide in favor of inoculation. With readers confused by contradictory arguments and unnerved by the risks of inoculation, debate continued for decades.

## FRENCH PHILOSOPHES AND INOCULATION

Even among *gens de lettres*, inoculation proved divisive. Julien Offray de La Mettrie, a philosophe/physician, opposed the procedure, whereas the philosophes' erstwhile enemy Élie Catherine Freron supported it.[13] Pro-inoculation philosophes abounded, however. To them, inoculation represented an enlightened discovery *par excellence*. Inoculation provides yet another example of how fractured the Enlightenment was. It was not a concrete program but rather a series of shifting debates and alliances. Nor did these debates tend to be mild. Those who criticized inoculation, most savants insisted, were ignorant, misinformed, or irrationally religious. Inoculationists were determined to ensure that the procedure would spread and that reason would triumph over doubt and superstition. As ever, philosophes had confidence in their own values and denigrated those of others.

Voltaire, accomplished as he was in such rhetoric, was among the first to pick up the banner of inoculation. He included a discussion of the technique in his *Letters Concerning the English Nation*, published in 1733; this constituted one of the earliest efforts to propagandize inoculation in France. Inoculation could reduce smallpox-related fatalities and disfigurements by the thousands, he claimed, just as it already had in England. If the French would inoculate, many lives would be saved: "The Duke de Villequire . . . would not have been cut off in the Flower of his Age. The Prince of Soubise, happy in the finest Flush of Health, would not have been snatch'd away at five and twenty; nor would the Dauphin . . . have been laid in his Grave in his fiftieth Year." Likewise, "twenty thousand Persons whom the Small-Pox swept away in Paris in 1723, would have been alive at this time."[14] Thus while famous Englishmen survived and prospered, their French counterparts fell victim to disease; whereas thousands of Britons were preserved, thousands of French men and women sickened and died. Inoculation would change all of that.

Voltaire rarely pulled rhetorical punches. In his telling, he fought on behalf of reason versus ignorance and superstition, toleration versus oppression, progress versus decline. His tendency to see the inoculation debates as a battle between light and dark only amplified as the French remained hesitant to inoculate. Frustrated but still hopeful, he wrote to Joseph Michel Antoine Servan in 1766 that the French would, sooner or later, stop resisting change. He listed, in his discussion of enlightened knowledge, "the circulation of the blood, gravity, the refrangibility of light, and inoculation."[15] Voltaire thus grouped inoculation with the ideas of William Harvey and Isaac Newton, which he understood as important victories in the culture war between reason and superstition. He hoped to soon list inoculation as a victory of a similar magnitude.

Voltaire's fellow philosophes generally matched his support of inoculation. Those involved with the *Encyclopédie* spread the gospel with zeal. In his article on "smallpox grafting," for example, Diderot described the operation as "the most wonderful discovery yet made in medicine for the conservation of lives." Jaucourt's piece "Visage" echoed this language: "Inoculation . . . is the most wonderful and most useful discovery in all of medicine." Damilaville, in his article "Population," likewise praised artificial smallpox as a "wise method . . . whose happy effects the nations of the world are already enjoying." The article "Inoculation" by an anonymous author further praised the technique that would one day, however slowly, win the support of the French: "Ignorance, superstition, prejudice, fanaticism, and a lack of regard for the public good will slow its march & they will fight us at every step. But after ages of struggle, we will finally have our moment of triumph."[16] Many philosophes thus understood inoculation to be a hallmark of "an enlightened century" that would have wondrous effects on health and prosperity, so long as its progress was not halted by doubt and superstition.[17]

In praising the amazing potential of inoculation, philosophes appealed to their fellow citizens. As the eighteenth century progressed, *gens de lettres* felt increasingly confident in the wisdom of "the public," and they beseeched it directly. Inoculation became one of many topics debated in the public sphere of eighteenth-century France. In pamphlets, in newspaper articles, and in books, savants made their case. Urban readers had ready access to these texts and many others (even forbidden books) through a dense network of cafés and bookstores, where they eagerly debated various political, social, and cultural issues.[18] The inoculation debate blossomed at a particularly fertile time for public discussion.

Treatises and newspaper articles were especially influential on this matter because of political and medical indecision. For the first few decades after inoculation's introduction in France, individual doctors, savants, and interested authors weighed in, but there were few coordinated official responses. Not until 1763 did the Parlement of Paris issue an edict stopping inoculation in the cities under its jurisdiction and ordering the Paris medical and theological faculties to render judgments. This edict was not intended to quash inoculation—interested patients could travel to the countryside and be inoculated there—but to clarify the risks and benefits of the procedure for the public. The faculties remained mired in confusion and internal discord, however, and their deliberations in no way simplified the matter for the public. Public debate continued to flourish in the absence of a clear decision, with a flurry of texts arguing for and against inoculation.

The authors of these texts revealed themselves as masters of emotional manipulation. In addition to crafting rational arguments about the benefits and drawbacks of inoculation, they also evoked fear by alluding to the disastrous consequences that could befall those who disregarded their advice. For pro-inoculation writers, that meant demanding that readers consider the possibility of watching a loved one die slowly and painfully from natural smallpox. As Samuel-Auguste Tissot, a celebrated doctor, academician, and author of *Inoculation Justifiée* asked: Was it possible to imagine a crueler scene than "that of a tender wife, forced to distance herself from her husband during the time in which his needs are the greatest! Than poorly cared-for children, who might even die, all because their mother had feared, or because someone else had feared, had not dared, had not been able, to care for [inoculate] them!"[19] Savants like Tissot demanded that their readers tremble at the dangerous consequences of ignoring inoculation.

Clearly philosophes had grown confident in their own cultural and moral authority. In Voltaire's *Lettres Philosophiques*, he modeled himself as a brave defender of reason and truth doing battle against the forces of ignorance and superstition. He solidified this reputation through his involvement in the notorious Calas affair of the 1760s, when he defended the Protestant Calas family against charges that they had murdered their son for having recently converted to Catholicism. Voltaire launched a passionate print campaign and eventually saw the verdict reversed. This public success reinforced Voltaire's certainty that he and his fellow *gens de lettres* had the ability to influence radical cultural change.

Philosophes' influence did, in fact, grow in the second half of the eighteenth century. Many thinkers attempted to shape public opinion on topics such as religious toleration, education, science, and statecraft. The cause of religious toleration became increasingly popular; philosophes gained control of the academies; they witnessed the exile of the Jesuits, one of their political enemies. The 1760s represented a period of success. Yet despite their growing cultural authority, the philosophes struggled to convince the French public to inoculate their children.

Although pro-inoculation philosophes ultimately failed in their quest to make the procedure universal, their efforts merit consideration. Even failed attempts to influence the public reveal much about norms and assumptions. Moreover, the inoculation campaign showcased new persuasive tactics, especially savants' use of sentimental and domestic language to bolster their authority. The inoculation crusade thus reveals both the new power of the philosophes, for they featured themselves as men worthy of emulation, and the limitations of their influence, for much of the French population resisted inoculation.

## ENLIGHTENED CALCULATIONS, ENLIGHTENED LOVE

To overcome resistance to inoculation, men of letters experimented with new techniques of mathematical calculation and quantitative analyses of infectious disease. Most significantly, the mathematician Daniel Bernoulli developed the first mortality index for inoculation in 1760. This allowed him to quantify inoculation's benefits by specifying how much longer a treated individual would live when compared with an untreated individual. He derived this number by estimating how many infants would survive to adulthood if smallpox did not exist (which he assumed would be the case if all children were inoculated). He then compared that number with current mortality rates. His results were compelling. Bernoulli's mortality indices demonstrated the risk of dying of smallpox was 15 percent without inoculation but only 3 percent with the procedure.[20] These mathematical calculations gave a bird's-eye view of inoculation and showed that, in the aggregate, the procedure greatly benefited the population. With many concerned that France was in the midst of a depopulation crisis, this was a significant finding.

Bernoulli focused his analysis on the state, which would profit from a greater number of adults able to lead useful "civil lives."[21] Inoculation would increase the number of subjects in France, spurring population growth, tax revenue, and military expansion. Bernoulli imagined a new role for the state in managing public health. Previously, the state had only intervened in health crises, for the early modern French state had neither the inclination nor the ability to guide public health more generally. But inoculation opened up a new way of thinking about government intervention in medicine: the state could sponsor, even require, preventative treatment rather than waiting for an outbreak to strike.[22]

It was hard to argue with Bernoulli's conclusion that promoting inoculation was to the state's advantage. But was that the right question to ask? Not at all, said Jean le Rond d'Alembert in his "Thoughts on Inoculation." D'Alembert stressed that the decision to inoculate was highly idiosyncratic. Parents cared little about the state when making decisions about their young children. D'Alembert doubted that Bernoulli's statistics would have an impact: "I do not know if this difference in probability would be sufficient to justify the operation from the perspective of a father who has lost a son to inoculation; I doubt even further that this reasoning could console such a mother."[23] It was ridiculous, he argued, to assume that all individuals would interpret evidence in the same way or that everyone would come to the conclusion that inoculation was necessary. It was especially absurd to assume that a procedure that was beneficial for the state was *de facto* beneficial for individuals and their families; the interests of the state and the interests of the individual were not so easily conflated. Denis Diderot disagreed strongly with this interpretation, however, sniffing that he would have thought "a man more attentive to the general good, not just his reputation, would have kept this essay in his pocket . . . [for it] has given so much pleasure to the idiotic adversaries of inoculation."[24] For Diderot, inoculation was the unselfish, rational, and correct choice; all good citizens would elect to have the procedure. Like many of his contemporaries, he urged his fellow citizens to cater to the needs of society rather than their own particular interests.[25] By preserving countless lives and preventing outbreaks, inoculation would do exactly that.

Although Bernoulli had demonstrated that inoculation was less risky than smallpox itself, his tables also showed that children did still die as a result of the procedure. That fact gave many parents pause. Many could not bear the thought that their actions, however well intentioned, might kill their children. Intellectually fascinating though it was, Bernoulli's mortality index did little to persuade individual parents that they should inoculate their children. Men and women of letters needed another persuasive technique, preferably one that would address nervous parents directly.

One method involved drawing on the growing literature on "good" fathers in eighteenth-century France. Fathers had long wielded considerable legal power over their families, and in seventeenth-century France, this power was culturally linked to the absolutist monarchy. The father was king of his family; the king was a father to his people. By the 1750s, however, this image had lost some of its appeal. Rather than stressing a patriarch's unchecked power over his household, writers stressed fathers' love for and obligation to their families. A good father worked to ensure his children's health and happiness.[26] He did not lord his authority over them but instead strove to lovingly and reasonably guide his household, keeping in mind their wishes and future happiness. Patriarchs could still expect to influence their children's decisions—about marriage, careers, and other matters—but within certain boundaries. A bad father would heavy-handedly insist that his daughter marry a man she did not and could not love; a good father would work with his daughter to choose a prudent but affectionate match.[27] The ideal household remained patriarchal, but a softer sort of paternalism flourished within this framework.[28]

The archetype of the good father appealed to many in Enlightenment France and was especially prevalent in the literature on sentimental families. The good father ruled by love; his children affectionately obeyed. The figure cropped up in various milieux, including novels, plays, paintings, pamphlets, and natalist petitions. Louis XVI, touched by these new ideas, aspired to rule as a paternalist "good father." He made his love for his family, his concern for his subjects, and his sensitivity to human suffering key tenets of his image (although he soon discovered that being a good father did not spare him from being mocked for his lack of virility and inability to control his wayward wife).[29]

For this and other reasons, pro-inoculation writers gravitated toward examples of fathers inoculating their children. This tendency was rooted in gender biases, as some authors assumed that women would be unable to handle the stress of inoculating their children.[30] Distraught at putting their children in harm's way, unable to appreciate the long-term benefits of inoculation, these imagined women bore little resemblance to the many mothers who chose to inoculate their children. The fearful women seen in these texts instead matched prevailing assumptions about mothers' over-sensitivity, attachment to their children, and inability to appreciate the bigger picture. That fathers had the final word on the subject—for patriarchs could overrule matriarchs in this matter as in others—also explains the focus on stories of fathers inoculating their children. The good father seemingly had more influence over the fate of inoculation than the good mother.

Influenced by new ideals of paternal tenderness, pro-inoculation philosophes wrapped inoculation in the cloak of family love. A good father—one anxious to safeguard his beloved children's future—would necessarily inoculate them. Charles-Marie de la Condamine proved especially adept at working this angle. A well-respected savant, La Condamine was best

known for his expedition to the Amazon until he acquired a reputation
as France's premier advocate for inoculation. From the middle of the century until his death in 1774, he devoted himself to combating smallpox, a
disease that he described as "an instrument of death that strikes without
consideration for age, sex, rank, or climate; few families manage to avoid
paying the fatal tribute it demands."[31] His efforts reached their apex between 1754 and 1765, when he read three celebrated *mémoires* on inoculation to the Academy of Sciences; all three were subsequently published.

La Condamine's writings appealed to his readers' emotions. He devoted much space to discussing how proper paternal sentiments such as
love and prudence would lead fathers to inoculate their children. La Condamine joined the cultural conversation about good fathers, and added "inoculator" to the list of characteristics that marked a *bon père de famille*.
By linking inoculation with the highly popular ideal of the good father, he
boosted the procedure's image and infused it with sentimental warmth.
Rather than focusing on the benefits inoculation would offer to the state,
he instead wrote emotional narratives that counseled French parents on
how to best care for their children. He was a shrewd salesman who knew
how to read cultural trends. He harnessed the power of sentimental rhetoric and hitched inoculation to the ideal of the good father. He softened the
technique by associating it with tender paternalism. This was not a scary,
detached exercise, he seemed to say; instead, inoculation was a reasonable
choice made by loving parents.

One particularly striking aspect of La Condamine's memoirs was the
concept of *amour éclairé*, or enlightened love. In an imagined dialogue, a
skeptic queried: "Can one ever persuade a tender father to wound his only
son . . . to give him a disease that he might have avoided, and which might
kill him? However small this risk might be . . . should the father expose
him to this illness voluntarily?" La Condamine answered, "Yes, without
a doubt." If the father loved his son with the appropriate kind of affection,
then he would perceive only one true path: he must inoculate his children. "If he wants to save his children from an incomparably greater risk,"
La Condamine wrote, "if prejudice does not obscure the father's reason, if
he loves his son with an enlightened love, he should not hesitate to inoculate the child."[32] The message was clear: if the father loved his children
properly, if he loved them with "an enlightened love," he would inoculate
them at the earliest convenience.

"Enlightened love"—what a pairing. The concept speaks to the complex blend of reason and sentiment that characterized the French Enlightenment, for men as well as women.[33] In arguing that parents should love

their children with an enlightened love, La Condamine insisted that love and calculation need not be opposed. Loving his children in no way prevented a father from making rational decisions about their future. Tissot played with similar language in *Inoculation Justifiée*, in which he asserted that "all evidence demands inoculation. . . . Every sensible and sensitive man to whom Providence has given children must, if he truly loves them, have them inoculated."[34] Once again, here was an image of the good father as a reasonable man, well in control of his family, but with his child's best interests as his guiding light.

La Condamine did not hang his whole argument on love but also relied on quantitative evidence. "The risk of death from smallpox increases from the time of birth," he noted. "One of fourteen newborns die of the disease; one of eight of those aged one year; and I have calculated that one of seven will die from the age at which one normally inoculates."[35] Complications from inoculation, on the other hand, killed far fewer people. Because the odds tilted so heavily in favor of inoculation, he argued, it was only reasonable to brave the risks the procedure entailed rather than waiting to be attacked by the more virulent natural form of the disease.

In La Condamine's ideal society, therefore, a father would be motivated by love and reason to protect his children from death and disfigurement, however much it pained him to expose them to the risks of inoculation. Both intellect and affect would encourage him to inoculate his children: "clearly, his reason advises him to do so and his paternal tenderness demands that he reduce this risk as much as he is able."[36] Parents who loved their children in an enlightened fashion would draw on their strength and sentiments to make brave decisions.

Indeed, La Condamine claimed that the choice to inoculate was not unlike other decisions that parents made about their children. Fathers married off their daughters all the time, he reasoned, and thereby subjected them to the risks of pregnancy and childbirth. "If his daughter died in childbirth, would [her father] blame himself for her death?"[37] Inoculation was no different and was even, La Condamine insisted, less of a risk than the peril of pregnancy.

These were all sly arguments. La Condamine manipulated cultural trends, most especially sentimentalism, as a way to bolster the public appeal of inoculation. He used the language of feeling to mediate between the public and the medical community. He cashed in on the eighteenth-century fascination with good fathers. He filled his texts with examples to show that inoculation was a well-proven technique and that many doctors knew how to perform it. He crafted his memoirs with care, working to make them as persuasive as possible. The stories he chose to tell reveal the significance of sentiment and empiricism as pillars of intellectual and cultural authority. La Condamine wanted to appeal to as many readers as possible, to have a maximum impact on the public. Sentimental language and carefully chosen case studies helped him do that.

La Condamine's insistence on the compatibility of reason and senti-ment, love and calculation, was common in inoculation texts. Not even D'Alembert—one of Bernoulli's major critics—claimed sentiment and reason were diametrically opposed. Instead, D'Alembert argued that the possible outcomes of reasonable reflection and emotional leanings were multiple and unpredictable; it was impossible to know how a father would choose to act and why. Jean-Jacques Rousseau, on the other hand, stated the terms of the debate in baldly calculating terms. In Book 2 of his *Emile, ou de l'éducation*, he noted that inoculation "conforms more nearly to our practice by saving his life at an age when it is more precious and risk-ing it when it is less so, if one can even give the name of risk to a well-administered inoculation."[38] Rousseau thus began his discussion of inocu-lation in positive terms, but it is his detachment that is most surprising. He invoked neither parents' love for their children nor the potential grief they would suffer if their children were to be carried away by disease. He was unwavering in his assumption that older children or adults were more valuable than young children. His tone probably did little to encour-age adoption of the procedure, as did the fact that he ultimately decided that Nature, not a doctor, should decide if and when Emile contracted smallpox.

The fact that most of these writers (Rousseau notwithstanding) drew on both reason and emotion in making their cases is not surprising in light of recent work on French intellectual culture and sentimentalism. Many *gens de lettres* relied on emotional engagement, particular observa-tion, and physical experience. They based truth claims on a foundation of particular experience and emotional attachment, rather than deductive thought exercises and abstract "systems."[39] For a savant to appear coldly calculating or overly detached called his character into question and ren-dered his work suspect; a good empiricist would emphasize his emotional connection to those around him. Pro-inoculation texts relied on tropes of family love and emotion to make them relatable and persuasive. Love, tenderness, horror, and fear would motivate individuals to inoculate their children; their reason would enable them to perceive the procedure's benefits.

## PERSONAL POLITICS

As in other eighteenth-century campaigns, philosophes involved in the inoculation debates presented themselves as moral leaders. They sought to root their appeals to the public in a politics of authenticity, to eliminate distinctions between what they thought and how they lived. They lived their lives according to their philosophical principles, or at least they did in theory. They based their intellectual authority on their moral goodness and, when relevant, their personal experiences. Philosophes were not perfect, and many had shortcomings as husbands and fathers. But in print, they came across as living versions of the "good father" so popular in novels, plays, and pamphlets.

Accordingly, many philosophes peppered their writing with personal stories, including stories that featured *gens de lettres*. In so doing, they held themselves and their compatriots up as model philosophers, practitioners, *and* parents. These stories not only helped philosophes sell themselves as models of parental love but also helped them counter their critics' attacks on their characters.

Examples of parents successfully inoculating their children were intended to encourage others to follow in their footsteps. Such stories would also show that inoculation was practiced by individuals of sound reputation and personal virtue. Sometimes, La Condamine listed a bevy of examples to prove beyond a shadow of a doubt that the best and brightest had chosen to inoculate:

> Mme la comtesse de Walle, Mme la marquise de Villeroi, Mme la comtesse de Forcalquier, dared to have themselves inoculated. It was M. Tronchin who directed the operation for these last two, as well as for many others, during his trip to Paris. The most famous of these were the inoculation of M. Turgot ... [and] M. la marquis de Villequier, of the son of M. d'Héricourt, former Intendant of the galleys, of M. de Vernege ... & that of the eldest son of M. le duc d'Estissac. ... In the springtime of the same year M. Hosly alone inoculated Mme la comtesse Walle, Mlle Quanne, the two sons of M. le marquis de Gentil, & the following autumn he inoculated M. le marquis de Belzunce, aged fourteen years.[40]

That La Condamine would rely on individual examples is not in and of itself surprising. Personal anecdotes popped up frequently in medical treatises. Before the development of quantitative methods and modern mortality measures, case studies permitted writers to accumulate and organize evidence. Individual stories also provided a narrative frame for medical writing, which made texts comprehensible and persuasive.[41] They also helped associate new procedures with social elites, which informed the reader of the many royal, noble, and otherwise illustrious parents who had inoculated their children and presumably hastened its wider adoption.

Anchoring medical arguments in case studies allowed writers to engage readers by spinning tales full of touching emotional details. The facets of these case studies that La Condamine highlighted, as well as the kinds of stories that he found compelling, are revealing. For La Condamine, examples did not just testify to inoculation's efficacy but also spoke to the larger social vision of which inoculation was a part. He frequently described parent-inoculators as "tender" and made clear how much they loved their children; here were figures who basked in the glow of "enlightened love." The story of a worried mother waiting by her son's bedside as a doctor inoculated him would heighten suspense. Her relief at his recovery would be palpable, as would her joy that his future would no longer be darkened by fears of the pox. In the age of sentiment, tugging at a reader's heartstrings was a surefire persuasive technique.

La Condamine also peppered his text with references to famous doctors who had become skilled at inoculation. In addition to the much-beloved Tronchin, he also noted that "by 1756 M. Tissot, author of 'Inoculation Justified,' had directed forty-two inoculations in [Lausanne] without incident."[42] These remarks reinforced the major point of La Condamine's argument: both love (many "tender" parents were opting for the procedure) and reason (physicians had become very skilled at inoculation) encouraged inoculation. Such case studies had a further benefit of making smallpox inoculation seem European, as opposed to stressing its Ottoman roots. Case studies helped represent inoculation as local knowledge, not an ill-understood foreign whim.[43]

Case studies suffered from a fundamental weakness, however. For every successful inoculation, for every loving family thrilled by the results of the procedure, detractors could find examples of inoculations gone awry, of families devastated by loss. One of La Condamine's most adamant critics, a doctor named M. Gaullard, wrote about two young boys who had been inoculated and had at first done quite well. However, on "the fourteenth day after the infection . . . the younger of the two inoculated brothers fell into a drowsy state which slowly worsened and he became [increasingly] lethargic until he died on the twenty-third day after his inoculation."[44] Andrew Cantwell, another skeptic, packed his *Dissertation sur l'Inoculation* with sundry references to grieving families.[45] Terrifying stories like these could scare off hesitant parents.

When confronted with potentially damning anecdotes, pro-inoculation writers argued that the particulars of the case were not representative. In the Gaullard example discussed above, for example, La Condamine contested the facts of this case and argued that "his body was opened thirty hours later, without any appearance of decay . . . one found in the base of his skull a pooling of serum; this was ruled the immediate cause of his death."[46] Moreover, he argued that further investigation revealed that the child in question had fallen ill before his inoculation but that the child's governess had concealed his sickness from his mother for fear that she would be accused of negligence. As such, the child was a poor candidate for the procedure. Inoculation should not be blamed for his death.[47] La Condamine's vigorous defense of inoculation in light of this particular case reveals how important—and how precarious—case studies were as evidence for and against the procedure.

The contested nature of case studies brought the question of authority to the fore: Who should the public trust? Whose stories should they believe? Pro-inoculation *gens de lettres* took direct action to establish themselves as both credible and honorable. They constructed a powerful new persona that fused their intellectual and domestic roles by arguing that they were virtuous family men who had inoculated themselves or members of their families. They transformed the domestic sphere into a site for the public demonstration of knowledge. They held up themselves and their compatriots as models of domestic behavior, not just as philosophical advisors. In holding up their own families and those of fellow *gens de lettres* as case studies, philosophes built on existing storytelling traditions but added an intriguing new twist for the reader.

Burgeoning ideas of patriotism and civic virtue also shaped these texts. Writers of all stripes exhorted the French to love their *patrie* and to be selfless in their devotion to their country and their fellow citizens.[48] One author, Claude-Louis-Michel de Sacy, nicely summed up this iteration of civic virtue: "I call Citizen any man who cherishes his *patrie*, who prefers the common good to his own well-being, and who is prepared to sacrifice to the society in which he lives his tranquility, his liberty, his life."[49] Patriotism did not fall under the purview of any single social order; nearly anyone could be a good citizen. Inspired by this rhetoric, supporters of inoculation represented the procedure as a patriotic act: it was a small personal risk that could yield great benefits for the nation and for one's fellow citizens. Nor was this the only way that the inoculation debate was bathed in the glow of patriotic love. In the last decades of the eighteenth century, physicians represented themselves as consummate enlightened, patriotic citizens. They claimed to serve the public good, not just a few wealthy clients. They disseminated the latest and most effective medical knowledge to the public and helped the poor and sick at great personal risk. They worked themselves to the bone in their efforts to improve humanity.[50] At multiple points, therefore, the inoculation debate intersected with the cult of patriotism booming in eighteenth-century France.

When men of letters modeled themselves as selfless patriots, as virtuous fathers acting for the good of their children, they borrowed a page from the nobility's playbook. Nobles had long represented themselves as men and women of virtue and merit who selflessly served king and nation. Men of letters did not replace nobles as presumed moral leaders; inoculation texts, for example, frequently dwelled on examples of noble inoculators. Nobles remained prominent figures in French society, and the role of the nobility prompted much discussion. Nobles were at the forefront of developing new ideas of patriotism and civic virtue, qualities that they insisted the nobility displayed but which were, at some level, available to a wide swath of the public.[51] Inoculation treatises, with their juxtaposition of nobles and men of letters as figures worthy of emulation, speak to the continued prominence of the nobility and the ways in which noble values were being adapted by nonnobles.

In their efforts to portray themselves as virtuous, publicly engaged, and patriotic, French philosophes were inspired by early examples of learned men and women inoculating their children. French philosophes were not the first to use their offspring as test subjects. The most prominent example cited, one that appeared in nearly every text on inoculation, was that of Lady Mary Wortley Montagu, the woman of letters and wife to the Turkish ambassador who introduced inoculation to England. Montagu's decision to inoculate her own son was a near-constant feature in these texts, and her bravery and willingness to make an example of her beloved child was often identified as the key factor explaining inoculation's early adoption in England. Montagu advertised her success to the Princess of Wales, who then tested the procedure on four condemned prisoners. Once she felt confident that the procedure was safe, she had her own children inoculated. Many followed suit. Voltaire praised Montagu in his *Letters Concerning the English Nation*; Tissot lauded her in his *Inoculation Justifiée*. French philosophes had their doubts about French women and their fortitude in making the decision to inoculate—this despite the fact that women often advocated inoculation—but the example of Montagu and other strong-willed women suggests that feminine weakness was not perceived to be universal.[52]

Montagu had a strong sense of her own patriotism and bravery in inoculating her child. In the same letter in which she described the procedure and stated her plans to have her son undergo the procedure, she added: "I am patriot enough to take pains to bring this useful invention into fashion in England, and I should not fail to write to some of our doctors very particularly about it." Montagu doubted, however, that many doctors would embrace the procedure; too much of their income came from smallpox patients. She told her friend that upon her return to England, she might, "however, have courage to war with them." Confident in the procedure's benefits, sure that others would "admire the heroism in [her] heart," Montagu made ready to fight for inoculation and, by extension, the health of her nation.[53]

In these descriptions, Montagu represented the ideal enlightened mother: loving, patriotic, learned, and committed to the well-being of her children and the education of her compatriots.[54] Her curiosity and brilliance enabled her to recognize the benefits of inoculation and her love for her son motivated her to protect him from the ravages of the disease. As Tissot wrote, "Her tenderness for the son whom she had born advised her to have him inoculated. The operation was a success, despite the ominous predictions of her chaplain, who wanted to invalidate, through a thousand arguments, a truth proven by a thousand facts. The child recovered very well."[55] La Condamine likewise praised Montagu, noting that this "woman famous for her wit had the courage to have inoculated in Constantinople, by her surgeon, her only son, six years in age, and then her daughter, upon her return to England, where this example was followed by many persons of distinction."[56] Montagu offered French men and women a model they would do well to follow: learned and maternal, she had bravely inoculated her own children. The results not only spared her family from future grief but also galvanized her English compatriots to inoculate their own children.

At nearly the same time, on the other side of the Atlantic, the eminent Puritan minister Cotton Mather learned about inoculation from his slave Onesimus, who had himself been inoculated in Africa.[57] Mather—who had long feared smallpox and the devastating consequences it might wreak on his family—embraced the procedure with enthusiasm. With the support of like-minded physicians, he urged his neighbors to inoculate themselves and their children. Mather's efforts landed him in considerable trouble. Many Puritans balked at the procedure—how saintly could it be to deliberately infect your family?—and suspected that "ingrafting" offered little protection against smallpox and instead caused serious complications. The minister's endorsement stirred up such controversy that someone even threw a bomb into his home, complete with a threatening note about inoculation.[58]

Mather nevertheless remained convinced that inoculation was a sound choice, and he worried that two of his younger children—not yet exposed to natural smallpox—might fall victim to the disease. Mather very much wished to inoculate these children (Elizabeth, 16, and Sammy, 14), for doing so would protect them. Mather may well have had other motives: successfully inoculating his children would, perhaps, help the public see the procedure in a new light. In the end, fearful that openly inoculating his children would provoke more outrage and attacks, Mather opted to inoculate his son Sammy in secret, presumably with the intention of publicizing his success after the fact. Unfortunately for Sammy, the inoculation followed an irregular course. He developed an unusually high number of pocks and suffered a worryingly high fever. His father feared that the inoculation had not been properly performed or that Sammy had contracted natural smallpox beforehand. For days, it was not clear if Sammy would live or die. He eventually recovered.[59] Mather's experience underscores how complicated the decision to inoculate could be. He wished to protect his son from harm, but he also hoped to make an example of him. Inoculating your own children was not a harmless decision but rather a flamboyant move that carried considerable risks.

English physicians who inoculated their own children in the 1720s and 1730s also played up their personal experiences. They mentioned in their correspondence and publications that they had submitted their children to the procedure. In so doing, they argued that the public should trust them. They had themselves deliberated over this moral and medical dilemma, and they had decided in favor of inoculation. Other parents should do the same.

French thinkers drew further inspiration from the tradition of savants experimenting on their own bodies, if not those of their children.[60] Curious about the effect of icy water on the human body? Jump into the Seine on a cold morning. Anxious to advance the surgical treatment of hernias? Volunteer as a test subject for a new operation.[61] Although such experiments seem reckless, they were celebrated as proof of a savant's intellectual engagement. Savants who risked their bodies acted bravely and put the interests of science above their personal comfort or even their lives. Self-experimentation proved their usefulness and virtue.

Experimenting on children remained a different story. Risking one's own life seemed brave, but risking a child's life might come across as foolish or unfeeling. Philosophes insisted that the opposite was true: they were rational enough to perceive the risks the procedure entailed and to inoculate in spite of it, and they loved their children enough that they would take care to preserve their future health. In ostentatiously putting their own children at risk, philosophes claimed that they were brave enough to practice an operation that frightened lesser parents. They held themselves up as examples of enlightened love: parents who loved their children, appreciated the benefits of inoculation, and took all necessary steps to protect their children from future harm.

Personal experiments of this sort happened alongside other, larger-scale experiments, including the inoculations of orphans, imprisoned criminals, and slaves. These groups might seem to have little in common, but in each case the state felt entitled to experiment with their lives. Orphans were wards of the state; criminals condemned to death had been separated from the body politic; slaves were property whose masters insisted they were less than fully human. If any of these individuals died as a result of the experiment, testers reasoned that few people would be troubled by the loss. Instead, demonstrating the efficacy of a medical procedure would, from the experimenters' perspective, allow dependents to make themselves useful. Observers saw these experiments as low risk: Who would mourn the loss of criminals already condemned to death? Objectified, ignored, or worse, these vulnerable populations were at the mercy of the state, and officials did not hesitate to expose them to diseases and cures.[62] Such experiments occurred regularly and provided empirical evidence that doctors used to lobby for inoculation.

Experimenting on one's own children served a different epistemological purpose, however. Here, the point was that philosophes loved their children deeply. Their attachment to their families, rather than their detachment from undesirable populations, imbued domestic experiments with special significance. They valued the lives of their children very much, and that infused their inoculations with drama and sensibility. At the same time, fathers could demonstrate their reason and understanding: they perceived inoculation to be the most rational and beneficial action. These inoculations commanded the readers' attention and valorized the savant as willing to risk his child's safety for the child's own sake *and* for the sake of the country.

For example, in 1766, the naturalist Philibert Gueneau de Montbeillard and his wife resolved to make their family a leading light for the cause of inoculation. Monsieur de Montbeillard traveled to Paris, observed famous inoculators in action, returned home to his country estate, and inoculated his only son with "the trembling hand of a father."[63] This inoculation—which Buffon noted required much courage—was a success.[64] Montbeillard then presented a memoir on inoculation at the Academy of Dijon, in which he drew attention to his actions as significant proof. "I desired, for the good of my country," he wrote, "that the example of a father inoculating his only son should become so standard that it was hardly worth noting." In addition to performing a patriotic service, Montbeillard stressed that he had acted in his son's best interest. He had given the matter much thought, but ultimately reasoned that because "the risk of waiting for [natural] smallpox was much greater than that of preventing it through inoculation, I saw my duty and I fulfilled it." His self-description mirrored La Condamine's discussion of enlightened love: "My every desire for prudence joined with the pangs of paternal love to lead me to examine the facts, to weigh probabilities, and to follow courageously the path that seemed best for my child, even if it was difficult for the father."[65] Although Montbeillard included the occasional modest aside, he brandished his personal experience, paternal tenderness, and intellectual acumen with zeal. This was how an enlightened father must act, he seemed to say. Diderot agreed, writing to Montbeillard that he was a model father.[66] Driving home the links between philosophy, medicine, and parenthood, his biographer Louis-Pierre Manuel noted that Montbeillard "wrote [his memoir on inoculation] with the same hand with which he had just inoculated his son."[67]

This ostensibly private event served an important public function. It armed Montbeillard with empirical proof that the procedure worked and that he was a rational but sensitive father who knew how to take care of his child. Using fashionable sentimental language, he portrayed himself as deeply attached to his son. He, like other sentimental savants, discovered that the language of feeling could be deployed in intellectual debates. Moreover, he could portray himself as a rational and curious man eager to learn more about inoculation and, once he had perceived its clear advantages, to try it out on his son. Montbeillard claimed to be a good father *and* a good savant, representing the two as harmonious.

But was it really so easy to balance the demands of father and philosophe? Montbeillard's striking discussion of his son's inoculation suggests that father-inoculators had complex motives. He clearly felt attached to his child, and probably had good intentions when he inoculated him. Yet Montbeillard also felt determined to make his son useful to the nation and to his father. Unlike D'Alembert, Montbeillard operated under the assumption that what was best for the nation was best for the boy. In this case, it worked out: the young M. de Montbeillard pulled through his inoculation just fine. It is nevertheless easy to imagine other scenarios that would have shown that public and personal interest did not always align. Moreover, Montbeillard went about the inoculation in a flamboyant way. Despite his lack of medical training, he chose to inoculate his son himself. He did not rely on a seasoned practitioner, as most inoculation treatises recommended. He then played up this choice in his memoir, which suggests that his personal ambitions shaped his actions just as much as his paternal tenderness. La Condamine had, after all, elevated his position in learned society as a result of his interventions in inoculation, and Montbeillard may have hoped for a similar result. He had only been a member of the Academy of Dijon for five years when he read his memoir; presumably, his flashy speech helped him make a name for himself. Savants like Montbeillard had many, not necessarily compatible, reasons for inoculating their children. Montbeillard's sentimental language may well have reflected his love for his son, but it served a persuasive function here. It enabled him to come across as rational and sentimental, all while raising his intellectual profile.

Another, more famous figure was Théodore Tronchin, a Swiss doctor well-beloved by philosophes and social elites. Tronchin was one of the leading inoculators in France, and pro-inoculation writers often mentioned his skill and success rate as a way to assuage anxious parents that the procedure was safe if performed by a gifted physician. Tronchin launched into hypercelebrity when he inoculated the children of the Duc d'Orléans in 1756. His own life also made good copy. Pro-inoculation writers underscored his courage and foresight in inoculating his children. La Condamine wrote: "In 1748, doctor Tronchin of Geneva, inspector of the medical college in Amsterdam, having nearly lost one of his sons to natural smallpox, took the part of inoculating his eldest; this was the first inoculation performed in Holland." This private decision had public ramifications because "it was followed by nine others that M. Tronchin directed; two years after that he recommended the practice in Geneva, his homeland. Thus it was that in 1750 that that republic, where the arts have flourished and where the zeal for the public good is a virtue common to all citizens, adopted the practice of inoculation."[68] Tronchin did not keep quiet about his son's inoculation. Instead, he used his success to influence others.

La Condamine was not the only one to use Tronchin as a key example. He appears in the two *Encyclopédie* articles titled "Inoculation," one penned by an anonymous author and the other written by Tronchin himself. The anonymous author noted, "In 1748, M. Tronchin, inspector of the medical college of Amsterdam, introduced *inoculation* in Holland, & began by practicing it on his own son. He recommended its usage in Geneva his homeland, where it was adopted in 1750." Tronchin's example inspired many others: "Two of the premier magistrates in that republic made examples of their daughters, aged sixteen years. Their fellow citizens imitated them, and since that time insertions have become more common."[69] In acting as a good father, Tronchin gave an example to many. His personal life enhanced his public expertise.

Tronchin agreed with his admirers. When he wrote an *Encyclopédie* article entitled "Inoculation," he highlighted his own son's inoculation. Writing about himself in the third person, he noted that "in 1748 . . . M. Tronchin . . . *inoculated* his eldest son in Amsterdam. He had lost his youngest son to the horrors of natural smallpox; his fear [that he might lose another] made up his mind [to inoculate his oldest]. This *inoculation* was the first seen in Christian Europe outside of the British isles." His son's inoculation inspired many others: "M. Tronchin did the same for nine other people with equal success. The next year M. Tronchin, having made a trip to Geneva, advised them to inoculate; his family gave the example, and others followed."[70]

In these tellings, Tronchin and his family courageously blazed a trail for others. As a result of his actions, much of the Genevan population inoculated their own children. Writers hoped that domestic experiences like these would affect public discourse and public actions. They saw Tronchin's decision to inoculate as more than admirable: it constituted a form of public service.

Tronchin and Gueneau de Montbeillard did not stand alone. La Condamine drew attention to other savant father-inoculators. His 1755 memoir informed readers that "M. Calendrini, a famous mathematician and a premier magistrate, gave an example using his son." He further noted, in his 1758 *mémoire*, that "in the spring, Madlle de Vaucanson, the daughter of the academician, proved that a nine-year-old child was capable of [surviving inoculation]."[71] Vaucanson was a celebrated engineer well known for his experiments with artificial life. His most famous achievement was the construction of a mechanical duck that seemed to achieve a new level of realism: it defecated. He was also a devoted father, and Condorcet noted in the savant's eulogy: "Having only one daughter who had lost her mother shortly after her birth, he wanted to direct her education himself, knowing of nothing that could be more dear."[72] A man with ambitious dreams of social advancement, Vaucanson may have seen his daughter's inoculation as a way to achieve greater celebrity for himself.[73] His daughter's well-being was not the only dividend the procedure paid. Like the other savants discussed here, he may have hoped that his care for his daughter would raise his public profile and finally gain him the academic prestige he believed he deserved.

From La Condamine's perspective, Calendrini and Vaucanson made interesting case studies because they would have been well known to the reading public. As he did with the other celebrated individuals included in his text, La Condamine highlighted the fact that well-known scholars had inoculated their own children and that such examples collectively proved that inoculation was safe and effective. Calendrini and Vaucanson were useful cases for La Condamine because they were men renowned for their learning and reason and therefore evidence that truly enlightened individuals would opt for inoculation. The procedure marked practitioners—to use the words of the Duchess of Saxe-Gotha—as "very much *à la mode* and free of prejudice."[74] Savants believed that it could only help the cause of inoculation to note that the nation's intellectual elite were choosing to submit to the procedure.

Jean Bernoulli likewise made flamboyant use of his family's inoculated bodies to lobby in favor of the procedure. As La Condamine observed, "The Messieurs Bernoulli . . . were not content to simply declare their support for inoculation" and hoped "to obtain initial proof for the faculties of medicine and theology in Basle." In an effort to secure such proof, "the younger of the two brothers, M. Jean Bernoulli and the only one married, wanted to contribute his example. In 1756, he had his two youngest sons inoculated; last year he inoculated the eldest"; he also had himself inoculated.[75] Bernoulli's desire to obtain proof that would sway the opinion of doctors and the medical faculty motivated his decision to inoculate his children and himself.

Bernoulli's efforts were successful. His personal experience with inoculation lent credence to his claims in favor of inoculation. Bernoulli himself was the picture of health, a compelling reminder that inoculation safeguarded physical beauty and strength. His eloquent speech was persuasive, La Condamine claimed, but "even more persuasive for his audience" was "the presence and the health of the speaker, on whose body the disease had not left a trace." Bernoulli's personal experience with inoculation, made apparent in his healthy countenance, "was a living proof that gave a new weight to his reasoning."[76] Bernoulli's own body, as well as the bodies of his children, lent gravitas to his speech.[77]

La Condamine used his family as public proof in a similar way. His eulogy, written by the Marquis de Condorcet, noted that he had inoculated his own family and that he "had the consolation of seeing his family give an example in his province, exposing ceaselessly to the public the success of inoculation . . . speaking always with a candor that persuaded and a warmth that he owed to his intimate conviction."[78] Condorcet saw La Condamine's decision to inoculate his own family as a key method that he used in convincing the public that they should inoculate their own loved ones. This, combined with La Condamine's "candor" and "warmth," helped constitute his authority.

When Montagu, La Condamine, Bernoulli, Tronchin, and Gueneau de Montbeillard highlighted their personal experiences with inoculation alongside those of other well-known savants, they emphasized that they and, by extension, their learned associates, were individuals whom the public could trust and emulate. Unlike their critics, who they said did not properly understand inoculation, these men of letters fully grasped how inoculation worked and had practiced it on their children. They asked their compatriots to follow their example. They invested the inoculations of their children with public significance and understood their actions as a form of social engagement.[79] Dwelling on stories of men of letters inoculating their children did more than provide empirical evidence for the reader: it invested *gens de lettres* and their allies with moral authority. By drawing attention to their personal lives, savants claimed they were more than learned. They were devoted mothers and fathers who had bravely inoculated their own children. They had been motivated by love of family and love of country, and their families were safer as a result. In stressing their personal experience with inoculation, therefore, savants did more than establish their empirical bona fides: they also represented themselves and their friends as loving parents who had made exactly the same decision they were asking the public to make.

These domestic experiments thus did more than provide philosophes with proof that they practiced what they preached (although that was important). They claimed to embody the blend of affect and reason that La Condamine called "enlightened love." They drew on popular notions of the good father and claimed that they fit the bill. In sum, *gens de lettres* staged a performance starring themselves as ideal parents who should be emulated.

The decision to inoculate had complex emotional underpinnings. In drawing attention to their actions, savants made two simultaneous claims about themselves: they loved their children and they were willing to put their own children at risk for the public good. Inoculation could cause complications, even death. In assuming these risks, philosophe-parents reaped two benefits if the inoculation succeeded: they protected their children and they burnished their reputation. Both their parental tenderness *and* their willingness to risk their children for the public good enhanced their public standing. Their intellectual authority drew on, paradoxically, their love for their children and their willingness to expose those children to danger.

That thinkers would use their families in this way, rather than staging public demonstrations with unrelated individuals, was a bold expansion of intellectual authority into the realm of the domestic. Savants turned the tools of natural inquiry onto the domestic sphere—their domestic sphere—for the first time. They argued that scientific reasoning could and should influence parents' decisions (as it did theirs). They turned both themselves and their children into "living proof" that validated their ideas.

Inoculation texts thus reveal new understandings of the role of savants in society. Thinkers depicted themselves and their fellow *gens de lettres* as individuals of empirical and moral authority who should be considered worthy of emulation. The inoculation debate was another forum in which *gens de lettres* represented themselves as true moral leaders of the nation. While other scholars have noted that men of letters aspired to a new public role at this time, they have done so primarily through analyses of academic speeches and *causes célèbres*.[80] The inoculation debates, however, reveal a new strand of this intellectual self-fashioning. Philosophes underscored the importance of their family lives, of their lived experiences, of their children's lived experiences. Inoculation, by virtue of the fact that it was a personal choice that every father had the authority to make, lent itself neatly to this form of self-fashioning.

In addition to revealing new facets of the public role of the philosophe, the fact that *gens de lettres* used their own families as case studies reveals important new aspects of scholarly authority. It was no longer sufficient for savants to simply have a well-reasoned argument; it had become increasingly necessary, in the age of sentiment, to demonstrate moral authority. Savants needed to show that they had the right motives by explaining themselves in the proper language and appealing to their readers' reason and sensibility. It also helped to claim personal experience, such as being an experienced inoculator or a parent who had inoculated his children. Although it was not essential to have personal experience in order to weigh in on the inoculation debate, it was one way that savants could claim public authority.

The inoculation of intellectual families thus served important epistemological functions: it permitted men of letters to represent themselves and their compatriots as loving family men who had considered their children's best interests before they chose to inoculate them. They could therefore be trusted to consider the public's best interests before they recommended inoculation on a broader scale. In this way, emphasizing savants' own experiences with inoculation mitigated potential concerns that they were unfeeling or were only interested in abstract thought exercises. Philosophes relied on case studies and personal experience as crucial evidence that they hoped would galvanize the French into inoculating their own families.

These texts also show that inoculation was tied up with notions of paternal virtue and civic duty. Those opposed to the practice did not contest these associations; instead, they adopted them. For example, Andrew Cantwell claimed that he had initially supported inoculation, even recommending the practice to others. But when it came to his own family, he could not bear to inoculate his children. That hesitation proved prescient, as Cantwell later turned against inoculation and argued that it was unsafe and ineffective. Here, as in pro-inoculation texts, Cantwell's love for his children and desire to protect them signaled that he was a good citizen and a good father whom his readers could trust.[81] Cantwell and other authors also sounded the alarm on inoculation because they worried it was dangerous for the public. Pro-inoculation writers in no way had a monopoly on sentimental and patriotic modes of writing.

That questions of personal morality and credibility were central to the inoculation debates is further apparent in the character attacks prevalent in both pro- and anti-inoculation texts. For example, La Condamine's adamant critic M. Gaullard thought that inoculation gave patients a disease other than smallpox, which made it a pointless risk. More to the point, Gaullard lampooned La Condamine as arrogant and ignorant, as immorally attempting to foist inoculation on the public. In an effort to expose La Condamine as a fraud, Gaullard challenged the philosophe to submit to a public inoculation, an experiment that the doctor felt certain would result in an infection even though La Condamine had already survived natural smallpox. If he did fall ill as a result of this inoculation, it would validate Gaullard's hypothesis that the procedure infected patients with a disease other than smallpox. If La Condamine developed symptoms of a disease, Gaullard would judge himself the victor in their debate.

Gaullard's challenge fits within the context of savants holding up their own bodies as proof of the validity of their ideas. If La Condamine had chosen to experiment on his body of his own accord, that decision would speak to his confidence in inoculation and would also orient him within the collective group of thinkers who drew on their embodied experience as scientific evidence. Gaullard's suggestion, however, had the air of a gauntlet thrown. If La Condamine did not experiment on himself, he would look like a charlatan and a coward. The same rhetoric that permitted savants to represent themselves as disciplined practitioners of the new science could cut the other way.

La Condamine took Gaullard's request as a personal insult. The latter showed "much bitterness, to the point of accusing me of bad faith; & not thinking of anything more than diverting readers' attention to contested facts." The very idea of challenging a rival to a public experiment such as this was repellent, and La Condamine reported that others had deemed it "ridiculous, outrageous, indecent even."[82] For Gaullard to imply that La Condamine did not dare inoculate himself, even as he urged others to do so, was to insinuate that he was misleading the public and that he was a man of "bad faith." Unsurprisingly, La Condamine did not appreciate this characterization.

In addition to critiquing Gaullard's motives, La Condamine also argued that the experiment was based on faulty principles. He scoffed that Gaullard's knowledge of inoculation was not as deep as the other man represented it to be. "M.G. claims that he will inoculate me in three different ways," La Condamine wrote, "that he calls the English, Chinese, and Turkish methods. The first method is, as we all know, by incision; the Chinese is that of inserting a bit of cotton coated with a powder made from dried pus taken from a smallpox patient." Both of these methods were well known to La Condamine, and he supported the method of incision. He had never heard of the third technique, the so-called Turkish method, which Gaullard claimed "consist[ed] . . . of making pills from smallpox pus, of which he reserves the composition for himself." As La Condamine had studied inoculation extensively but had never heard of this particular procedure, he was certain that it was an invention of Gaullard's. He had no interest in this "new exercise" but added that "I consented, I consent again to be inoculated: not by him, but in his presence; not with pills or powders but in the ordinary fashion."[83] La Condamine insisted that he was willing to make an example of himself, just not according to someone else's ill-informed whim.

This back-and-forth about a single experiment reveals the fundamental importance of a savant's personal reputation to his credibility and trustworthiness. Gaullard's request was shocking because he demanded La Condamine appear before a public audience and submit to inoculation by one of his opponents. The implication was that there was something untrustworthy about La Condamine, that he did not believe in the safety and efficacy of inoculation as fully as he claimed and so would hesitate to endure the procedure himself despite his willingness to expose innocent children to the same. This was a damning premise.

La Condamine responded to this proposal, potentially damaging as it was, by displaying his own authority vis-à-vis inoculation techniques: Gaullard's proposed "smallpox pill" was not practiced by any doctor in any society known to La Condamine. He asserted his honor and personal integrity, claiming he would be happy to demonstrate his commitment to inoculation by incision as long as it was performed by an impartial third party. He represented himself as more knowledgeable than Gaullard and more personally committed than the other had allowed.

The use of savants' personal lives as evidence of their moral and intellectual excellence as well as character attacks by and against philosophes points to the same central point: that questions of intellectual authority, trustworthiness, and moral goodness were fundamental to the inoculation debates. In attempting to convince the public to trust them, as opposed to those writers who lobbied against inoculation, pro-inoculation savants developed personalized forms of persuasion. They told moving stories about those who had reaped the benefits of the procedure and pathetic tales of those who had suffered the consequences of not inoculating. They privileged their own stories: their hopes and fears, their love for their children, their eagerness to protect their offspring from death and disfigurement, their willingness to use their beloved sons and daughters to educate the public. In their hands, inoculation became a daring act of paternal tenderness and enlightened reasoning. They hoped that many would follow their example.

## CONCLUSION

Inoculation excited much controversy in eighteenth-century France, and it continued to do so until Louis XVI—terrified by Louis XV's death from smallpox—inoculated himself alongside his brothers in 1775. These inoculations ultimately led the public to see the procedure as an effective way to prevent smallpox. But up until that point, there was much debate. In attempting to sway public opinion, pro-inoculation philosophes linked inoculation with cultural trends such as sentimentalism, paternalism, and patriotism. They developed a new social vision of an enlightened and healthy future, and the cornerstone of that social vision was the enlightened and loving family. Motivated by reason and sentiment, love for their children, and love for their neighbors, parents would decide in favor of inoculation. Philosophes thus envisioned the family as a vital source of enlightenment and parents as important agents of reform.

*Gens de lettres* saw their own families as especially important examples. They inoculated their own children and then used that personal experience as evidence that inoculation was indeed safe and effective. This intimate empiricism allowed men of letters to represent themselves and their colleagues as loving, rational fathers who acted in their families' best interest. Daring to inoculate their children, despite the risks it carried, made them more trustworthy. They were willing to put their families on the line and to accept that their beloved children might end up one of inoculation's rare victims. By caring for their children so deeply but, paradoxically, being willing to sacrifice them for the public good, they enhanced their public stature and intellectual authority.

Hence the family itself became an object of study in the Age of Enlightenment. Savants used the family as a sort of laboratory in which they could test and demonstrate their ideas for the public's benefit. Rather than sealing off their family lives as private, they threw back the curtains and laid bare their domestic practices as a public demonstration that they hoped would further the cause of enlightenment.

# CHAPTER FOUR

# Enlightening Children

When the celebrated anatomist Exupère Joseph Bertin died, the Marquis de Condorcet read his eulogy aloud to the Academy of Sciences. Bertin had spent his later years educating his children, Condorcet informed the crowd. He had "four children, whose education was a sweet, consoling occupation for him, his only duty which filled his days with pleasure and brought peace and calm to his soul, agitated by so many storms and torn apart by much misfortune."[1] Condorcet often included stories of devoted fathers, like this one, in his eulogies for members of the Academy of Sciences. Loving fathers seemed to abound amid academicians, at least in Condorcet's telling. Even those without biological offspring took part by lavishing attention on their protégés and treating them like family. Outside the academy, *femmes savantes* likewise worked to shape their children into worthy adults, and they studied pedagogical treatises and penned tracts of their own.

Many *gens de lettres* found educating their children a useful way to contemplate weighty philosophical issues such as happiness, morality, and sexual difference. Writers' descriptions of these educations often put a positive gloss on their experiences and skipped over many mundane details, but these sources still provide a tantalizing glance into the priorities and convictions of Enlightenment thinkers: what sorts of careers they expected their children to pursue, the differences between girls' and boys' educations, and what characteristics thinkers considered essential for the next generation. Influenced by their understanding of human nature as malleable, *gens de lettres* attempted to create an enlightened generation. Indeed, parents of all stripes—not just philosophes—aspired to raise their children according to ideals of nature, sociability, and sentiment.[2] They hoped to bring up a rational and loving generation, replete with citizens attentive to the dictates of nature and the needs of human society. One child at a time, they would transform the nation. In pedagogical texts, as in other discussions, the family served as the foundation of meaningful social reform.

As they did with inoculation, philosophes double-purposed the education of their children. They not only applied Enlightenment ideas to their children's upbringing but also seized the opportunity to, once again, depict themselves as virtuous citizens fully engaged in creating a moral society. They stressed their public utility and commitment to the nation, this time evidenced by their loving and rational care for their children. By using sentimental language and presenting themselves as loving parents committed to their children's happiness, *gens de lettres* seemed to be living up to the ideals of the sentimental savant. Their private virtues revealed them to be socially useful and trustworthy intellectuals. By depicting themselves in this fashion, thinkers were influenced by the ideals of the "good father" (the loving patriarch who engaged fully in his children's upbringing and affectionately guided them to adulthood, without so much as a hint of authoritarianism) and the "good mother" (the matriarch whose devotion to her children was all-consuming). These stock figures enjoyed great popularity in late eighteenth-century portraits, plays, and prescriptive literature.[3] Moreover, because they grounded their texts in personal experience, their writing had a strong empirical foundation. In this way, *gens de lettres* attracted readers' attention.

Writing about education thus shone a flattering spotlight on philosophe-parents—what loving and devoted citizens they were!—and articulated a method for raising the *next* generation of loving and devoted citizens. Philosophes leveraged their parenthood to craft a new moral vision for society, to present themselves as virtuous and publicly engaged, and to claim that they had indisputable intellectual authority. These issues of rhetoric and representation are at the heart of this chapter.

## DEBATING EIGHTEENTH-CENTURY EDUCATION

Looming over all educational writers was the work of John Locke. In his *Essay Concerning Human Understanding* (1690) and *Some Thoughts Concerning Education* (1692), Locke insisted that human beings came into the world without innate ideas or virtues. Babies were morally neutral. Their experiences from the moment of their birth shaped their character.[4] These formative experiences went beyond classroom instruction; what children ate, wore, and saw made equally strong impressions on their character.

Because every new experience imparted new knowledge to a child, potentially to dramatic effect, parents had to maintain a close watch on their children.

Within certain boundaries, parents and educators could shape a child's character. In particular, they could toughen a child's mind and body. For example, Locke recommended parents expose children to cold to make them impervious to chills that would bother weaker individuals. But toughening the body did not entail corporal punishment. If a child misbehaved, Locke counseled against spanking, whipping, or otherwise inflicting pain. Instead, he recommended a system of emotional rewards and punishments: parents should condition their children's behavior by expressing affection or disappointment. Locke's ideas proved hugely influential. Eager to present themselves as affectionate fathers who directed their children's education with care, eighteenth-century English fathers commissioned portraits in which they supervised their children writing on blank sheets of paper; the image recalled Locke's description of a child's mind as a *tabula rasa*, a blank slate.[5] Locke's texts also captivated the imaginations of French writers, who spent much of the eighteenth century expanding on Locke's ideas. Locke's sensationalism provided a path to mold children as philosophes saw fit, to form a generation of citizens and transform the nation from the bottom up. Indeed, eighteenth-century philosophers only grew more determined and more ambitious in their schemes to perfect society. Education provided the means to remake humanity, to find the sparkle in even a dull stone.[6]

French pedagogues agreed on common enemies as well as common inspiration. Bad mothers seemed to lurk everywhere, especially in the houses of dissolute aristocrats.[7] Allegedly more interested in social engagements than infant care, French women often relied on wet-nursing (when women, generally of middling or upper class status, sent their infants to live with nurses who breast-fed them). Some nurses neglected their charges; others were accused of corrupting children through their bad morals or manners. Many doctors and philosophes railed against the practice. Also subject to much criticism was the practice of swaddling (wrapping infants tightly in a blanket to immobilize them, keep them warm, and protect their growing bones). Swaddling constrained the child, in direct opposition to Locke's recommendations, and seemed to encourage dereliction of maternal duty. Education writers worried that swaddling made it too easy to stash a child somewhere, forgotten. Instead of wet-nursing and swaddling, thinkers demanded that mothers devote themselves full-time to nurturing their children and allowing them to play, learn, and develop self-control.[8]

These debates crossed into political and religious territory. Educational texts were, at heart, grounded in philosophies of human nature and visions of the ideal society. During the Revolution, education—especially an education founded on a sensationalist understanding of human nature—attracted attention as a means to transform the French into good republicans. Even before the heady days of revolutionary reform, however, education sat at a cross-section of political, religious, and cultural debates. Given that philosophes and conservative clerics found little to agree on here, education was a much-contested issue in the mid and late eighteenth century.

The debate heated up as the Jesuits, the chief educators in France, found themselves in trouble. The Jesuits' chief antagonists were not the philosophes but rather the Jansenists, a recalcitrant group of unorthodox Catholics. Jansenists and Jesuits diverged sharply in their theological principles, and the Jansenists—especially those holding office in the nation's law courts, the parlements—worked to undermine the Jesuits' position in France. They insinuated that the Jesuits were potentially treasonous, unpatriotic outsiders, more loyal to their order and to the pope than they were to France. They grasped every opportunity to highlight the Jesuits' mistakes, including bankruptcy, and eventually achieved their goal: the expulsion of the Jesuit order from France. Many anticlerical philosophes delighted at and, when possible, helped the Jansenists bring about the Jesuits' misfortune.[9] After parlementary maneuvering against the order, the Jesuits were removed from all teaching positions in 1762.[10]

This decision had major consequences, as the Jesuits were the predominant educators in France. They ran a vast network of secondary schools organized according to the order's course of study (the *Ratio Studiorum*); in 1761, there were 111 of these *collèges*.[11] Pupils focused first and foremost on learning to read, write, and speak Latin. Teachers evaluated their charges' progress by demanding that students memorize and recite passages. Students who misbehaved were subjected to corporal punishment; those who excelled were honored in elaborate ceremonies.[12] Many philosophes went through this system of schools, including Diderot and Voltaire, and they did not reflect positively on their experiences. Philosophes attacked the Jesuits for their perceived dogmatism and inflexibility, as well as their allegedly ineffective and harmful pedagogical techniques. This education, they believed, would stimulate the wrong kinds of sensations and result in the wrong kinds of adults. D'Alembert summed up the philosophes' derision when he wrote: "A young man, if he has spent his time wisely, leaves the college . . . with a very imperfect knowledge of a dead language and with precepts of rhetoric and principles of philosophy which he should endeavor to forget." These faults, coupled with warped or insufficient ideas about religion, left the young man ill prepared for his adult life and of little use to society or himself.[13]

The philosophes' disdain for Jesuit education cannot be boiled down to a straightforward disagreement about pedagogy. Many philosophes loathed religious orders: they charged monks and nuns with being socially useless, especially those who were cloistered, because they were celibate and secluded. Indeed, organized religion in general raised the hackles of many *gens de lettres*. The Jesuits could hardly have been accused of isolation—their missionary work led them around the world, and their schools made a direct contribution to French society—but the philosophes' writings about the Society of Jesus nevertheless featured more than a little anticlericalism. This helped drive philosophes' criticism of Jesuit pedagogy.

Also influential were new ideas of patriotism and citizenship, and especially the idea that education was a key process for transforming children into civically virtuous adults. Scores of French readers and writers hoped to renew the nation's patriotism, to encourage men, women, and children to work for the common good and to love their country and fellow citizens. Education could help achieve this goal, especially if it focused on teaching children about notably virtuous figures in France's past.

The Jesuits' downfall was appealing because it created a major hole in the French educational system and opened up a chance to create education practices more geared toward civic virtue. Many philosophes weighed in on the debate, but they found little common ground. Most writers shared Locke's sensationalism, enshrined childhood as a unique and formative period, and sought to develop secular virtue. They generally shied away from radical social reform and rejected universal equality as a desirable goal.[14] Yet they did not agree on the best methods to reach their goals. Some supported public education, arguing that the state should assume responsibility for molding citizens. Others advocated domestic education, insisting that parents were best equipped to guide their children. Some insisted on a radically egalitarian view of human nature, with all differences in intelligence ascribed to education; others argued that while education mattered, so too did inherited characteristics. To boot, there were a dizzying array of ideas about what and when a child should study, not to mention what differences there should be between girls' and boys' educations.

Among a crowded field of pedagogical writers, Rousseau stood out as the most widely read. His 1762 text *Emile, or on Education* proved remarkably popular, despite being banned and publicly burned for irreligious content. Rousseau's pedagogical program focused on preserving his imaginary charge's natural virtue and protecting him from the seductive but corrupting influence of society. He insisted (contra most philosophers of the time) that man was born good, that society had corrosive effects on physical strength and moral rectitude, and that parents must raise children apart from society if they wished to form them into autonomous, moral individuals. The fictional Emile grew up in the countryside, far from the unhealthy environment of the city, and devoted much of his early childhood to playing and exploring out of doors. Freed from the shackles of overly restrictive clothing and strengthened by his daily excursions, Emile developed virtue, curiosity, keen senses, and physical strength. Rousseau further insisted that his young charge not focus on formal study or religion in early childhood. Only later would Emile read widely and learn a trade. Once Emile became sufficiently mature, his tutor introduced him to his perfect match: Sophie. Rousseau revealed that Sophie's education had been quite unlike Emile's. Her mother taught her domestic skills and sought to preserve Sophie's naturally docile character. Passive Sophie complemented the active, engaged Emile. Indeed, the chapter on Sophie and Rousseau's other writing on women make clear that he considered public women highly disorderly; any society with a vocal and public contingent of women was necessarily corrupt.

Rousseau's work demonstrates how impossible it is to separate education from politics. *Emile* sought to accomplish something grander than tinkering with existing educational systems; Rousseau wanted to start from scratch. He provided readers with a roadmap to creating a new sort of man, a fully autonomous and virtuous individual. The key to this transformation was not the construction of new political institutions but rather a reorientation of family life around the practice of sentimental domesticity and individualized education. This was the radical heart of the book, and it explains why the text proved controversial with censors and alluring to readers.[15]

*Emile* was a runaway best seller. "Everything that *l'Ami* Jean-Jacques has written about the duties of husbands and wives, of mothers and fathers, has had a profound effect on me," wrote Jean Ranson, a merchant in La Rochelle.[16] A generation of men and women devoted themselves to parenting in the new fashion and sought to practice the ideas articulated in the novel. Moved by Rousseau's exhortations to follow nature's dictates and his insistence that mothers nurse their own children, many bourgeois and aristocratic women rejected wet-nurses and breast-fed their own children. Other, more drastic attempts to practice Rousseau's ideas included dripping hot wax on children and shooting pistols at them so that they might learn to face pain and fear without flinching.[17] As for Rousseau himself, he famously abandoned all five of his children to a foundling hospital, and so he could not boast of any hands-on experience with raising children.

Although it was very popular, Rousseau's work did not constitute the final word on education; his novel joined many contenders in the crowded field of pedagogical thought. Philosophes in general dreamed of producing useful citizens who could better serve their state and society, with "useful" being the key word. Du Marsais's *Encyclopédie* article "Education" nicely sums up this emphasis on utility:

> Children who come into the world, must form one day the society in which they will live. Their education is thus the most interesting subject, 1) for themselves, whom education must fashion such that they will be useful to that society, obtain its esteem, and find in it their well-being; 2) for their families, whom they must support and honor; 3) for the state itself, which must reap the fruits of the good education that the citizens that compose it receive.[18]

Indeed, the article mentions the word "useful" on several occasions, and shows that philosophes evaluated education according to a metric of utility. Being useful to oneself, one's family, and the state became the mark of a stellar education, and it was best accomplished through a focus on modern knowledge, languages, and vocations. An education focused on arcane, impractical knowledge fell short.[19] Professional competence and civic virtue outweighed introspection and erudition; philosophes were not interested in creating a new intellectual elite but focused instead on building a base of virtuous citizens.[20] Very few *gens de lettres* attempted to raise brilliant offspring who would impress contemporaries with their mastery of sophisticated knowledge. In this, they parted ways from their Renaissance predecessors. A stunning command of Latin and Greek, rhetoric, and ancient texts had marked young Renaissance thinkers as leading lights of their generation.

The desire to display one's children as better and brighter than their peers did not disappear in the eighteenth century, of course. Suzanne Necker, for example, worked hard to turn her daughter Germaine (the future Madame de Staël) into a prodigy. She schooled her daughter from the Bible when she was only two years old. Germaine's education eventually expanded to include the study of English, Latin, history, and geography. Necker instructed her daughter to write long essays on topics such as

"What is the best system of government?"[21] She then displayed the fruits of her labors at her salon, where she would lead young Germaine into her circle of friends. The child sat on a stool by her mother's feet, where she listened silently until she was called upon to dazzle guests with her elegant speech and ability to recite Latin passages on command.[22]

Most men and women of letters chose a different route, however, and attempted to form their children into exceptionally good incarnations of everyday citizens rather than extraordinary geniuses. They worked to teach mastery of accessible virtues and skills rather than more unusual or esoteric abilities. Self-control, selflessness, and the ability to perform useful tasks were the guiding lights of many Enlightenment educations.[23] Meritocratic ideas still mattered, of course. Philosophes tailored their pedagogical plans to develop unique talents and encouraged competition and emulation, which they believed would spur children to more virtuous lives.[24] But where education was concerned, philosophes tended not to think in radical terms. They did not encourage children to ignore the rules of propriety or to defy the strictures of social status. For them, an enlightened education was one rooted in sensationalist ideas, largely geared toward secular ideas of civic virtue. Their political vision entailed creating a newly moral and patriotic society, but gradually and without rejecting everything that had gone before.

The path to virtuous citizenship looked different for boys and girls, as gender was the key organizing principle of education in the eighteenth century.[25] Civic virtue and social utility were themselves gendered concepts, and so were the educations designed to foster them. For boys, a useful education generally meant learning a trade and preparing for public life. For girls, usefulness revolved around marriage and motherhood. Although girls were ultimately destined for domestic roles, they were essential for developing the nation's civic virtue. Women created and nurtured social bonds and personal morality. If wives and mothers were insipid, immoral, or stupid, the whole nation would decline.[26] A proper education would develop their minds, bodies, and morals and would ensure their happiness and their family's virtue. Philosophes raising daughters nearly all believed marriage and motherhood to be their daughters' destiny. They did, however, play with different ideas on how to best prepare them for these roles. Marked by an unwillingness to rethink the social hierarchy *and* a desire to find new ways to develop civic virtue, girls' education was "at once progressive and traditionalist," to borrow Lesley Walker's useful phrase.[27]

A common emphasis on civic virtue did not mean that every education

should be the same. Eighteenth-century thinkers insisted that an education must conform to a child's individual talents and social position. With
a full range of talents and inclinations, each generation would best serve
the nation; France would have citizens qualified in many areas. Individu-
alism was not at odds with the common good. Instead, a diverse public
would result in a stronger society, just as nature intended.[28]

Pedagogical texts centered on themes of gender, social utility, and
social reform might seem unlikely to make today's best-seller lists, but
many of these works appealed to a wide audience in the eighteenth century. Readers were enticed by the stories, dialogue, and heavy dose of sentiment that filled works on education. Authors aspired to entertain their
readers and presented vivid and accessible education plans to the public,
which they often staged inside real or fictive homes. This personal touch
grounded the text in specifics and made it easy to incorporate fashionable
outbursts of emotion. Women writers popularized this form in France.
By embracing maternity, the most laudable and "natural" of all women's
roles, they could make their philosophical writing more feminine and less
inflammatory. Male writers borrowed their techniques, according themselves and their children starring roles in their texts.[29] By depicting themselves as parents devoted to their children's education, men and women of
letters represented themselves as paragons of sensibility who loved their
children. And by emphasizing the care they took to make sure those children became useful citizens, they stressed their civic bona fides. Men and
women could thus use a sentimentalized rhetoric of personal experience
to draw attention to their pedagogical ideas and depict themselves as virtuous, loving, and learned.

## CASE STUDIES

### ÉMILIE DU CHÂTELET

Émilie Du Châtelet was once a shadowy figure, famed only for her romantic relationship with Voltaire, but historians have worked to restore her
reputation as a philosophe. She produced an array of publications, including her celebrated translation of Isaac Newton's *Principia Mathematica*.
Du Châtelet also invested considerable energy in raising her two children.
She used her maternity to craft an image of herself as a virtuous savant, as
opposed to an overly ambitious, unfeminine wannabe.

Her daughter, Gabrielle-Pauline, was born in 1726 and her son, Florent-
Louis, followed in 1727. Du Châtelet especially focused on preparing her

son for greatness. She never doubted that he offered his tutors "a pretty soul to develop."[30] In contemplating her son's upbringing, Du Châtelet drew considerable inspiration from her own interests. She encouraged her son to delve into natural philosophy, a subject freighted with moral and intellectual value. Although studies of nature had once paled in comparison with studies of religion, by the 1730s, natural philosophy had gained prestige, and Du Châtelet wanted her son to be up to date on the latest findings.[31] Studying nature would sharpen his intellect and reveal to him the impressive order of the cosmos.

Du Châtelet's education of her son took place against the backdrop of a larger debate of how nobles could best develop patriotism and merit. These questions interested Du Châtelet, and she thought carefully about how to raise her son. She wanted to form him into a leader who would serve his nation. She ultimately decided that the traditional path remained the best: the young monsieur should train to be a military officer. Less conventionally, Du Châtelet decided that he should be an especially learned sort of officer, one whose usefulness would in part be determined by his keen understanding of philosophy.

She even wrote a text to introduce her son to natural philosophy: the *Institutions de Physique* (*Foundations of Physics*, 1740), an overview of recent research in which she attempted to marry Leibnizian and Newtonian physics.[32] All parents should encourage such study, she opined. "I have always thought that man's most sacred duty is to give their children an education that would not cause them to regret their youth," she asserted. "[It is] perhaps the only time of your life that you can devote to the study of nature," which is "the key to all discoveries." It was at this age, the "dawn of his reason," that his mind was sufficiently "flexible" to take on the "punishing" study of nature. Indeed, Du Châtelet glorified the pursuit of natural philosophy to such an extent that she insisted that "the research of truth is the only thing over which the love of your country should not prevail."[33] She depicted natural philosophy as fundamental to her son's education, with the implication that other parents should do the same. Florent-Louis's education was rooted in nature, like many other Enlightenment educations, but with greater rigor than was the norm.

Du Châtelet plainly wanted her son to appear "enlightened," in the sense that he would display a certain amount of erudition. She also wanted to cultivate his patriotism and to instill a strong sense of duty to his king and nation. She shared this concern with other nobles, but she was unusual in that she decided studying natural philosophy would make him a better, more useful, more virtuous aristocrat. He would become a new

standard bearer for the nobility and noble service to the *patrie*. Traditionally, members of the sword nobility did not encourage the advanced study
of languages, science, and literature unless they intended the child to pursue a clerical career.[34] Science and literature seemed to bear little relation
to the nobility's traditional *raison d'être* of military and political service,
and some noble parents even feared that an ambitious education would
distract and weaken the child.[35] Du Châtelet dismissed such concerns
and designed a philosophical education that would make her son more,
not less, prepared to fulfill his noble duties. Learning about science would
protect him from "the ignorance which is all too common among those of
your rank, which is always a defect in excess and a merit in its absence."[36]
The study of nature would provide her son with an invaluable guide for his
own reason and conduct, and would thereby aid him in all pursuits. Most
of his generation would lack such an education and so Florent-Louis would
stand out as an especially meritorious individual.

"Merit" was a key term for the French nobility, the word they used
most often to describe their service to king and country. Earlier eras had
defined noble merit as generosity and self-sacrifice, but eighteenth-century
nobles linked merit with talent.[37] In cultivating merit through education,
the marquise operated under the same assumptions that guided her fellow
nobles. Because her son would be the most educated and the most skilled
in various disciplines, his merit would be readily apparent to his peers.
Although she held her son to a higher standard of erudition than did most
noble parents, her goal was the same: to make her son stand out as an especially good servant of the *patrie*.

In contemplating how education could mold her son into the ideal noble citizen, Du Châtelet anticipated the flurry of texts published after the
closing of Jesuit colleges in 1763. The departure of the Jesuits created a
huge hole in the educational system, and many nobles wished to make the
most of the opportunity created by their absence. Eager to transform noble
children into patriotic individuals, a host of authors argued that education
must be reconfigured to train young children to be moral citizens. Out
with Latin and ultramontanism; in with French history and patriotism.
As was the case with philosophes interested in education, nobles proposed
a range of educational plans. But on one major point they agreed: noble
children must learn to embody the selflessness, quiet merit, and tireless
devotion to duty that characterized the good citizen.[38] Du Châtelet may
not have made this point as forcefully as later authors, but she too wanted
to find a way to instill these values in her child.

When Florent-Louis was a teenager, his parents settled on a military career for him. This was a career choice imbued with noble tradition and classical merit. Du Châtelet remained vigilant in her efforts to ensure his success, appealing to her longtime friend the Marquis d'Argenson to help her son find an important post. After he acquired a military office, the Duc du Châtelet-Lomont (the title that Florent-Louis inherited) became part of the military reform movement that had gathered momentum in the wake of France's crushing defeat in the Seven Years' War.[39] Du Châtelet-Lomont became one of the most avid military reformers, working to make armies more orderly, disciplined, and meritocratic. Reformers sought to slash the number of show regiments, eliminate the sale of military offices, centralize recruiting, and make promotions more merit-based. Du Châtelet worked in accord with such ideals for most of his career, agitating for a more meritocratic system of recruitment and promotion.[40] Although Du Châtelet came to an ignominious end—becoming a scapegoat who had allegedly eroded loyalty among the French Guards during July 1789 and thereby "caused" the fall of the Bastille—he had enjoyed professional success up to that point.[41] His efforts to promote military reform were very much in line with the Enlightenment ideals of order, natural law, and merit. His study of the "new physics" and his mother's ideals had much influenced him. His mother, who died in 1749, saw only the beginning of his military career. Undoubtedly, however, she would have smiled at his successes and congratulated herself for her part in them. She had raised a leader who was pulling France into a new era.

Florent-Louis was not the only beneficiary of his education, for Du Châtelet worked to bolster her reputation at the same time that she instructed her son. At this point in her career, she had already sustained attacks that she was unlearned and arrogant. She crafted the *Institutions de Physique* to answer such charges. To stave off charges of arrogance, she insisted that the book "was created exclusively for the education of my only son, whom I love with an extreme tenderness; I believed that I could not give him greater proof of this than by making him less ignorant than the average young person."[42] She also claimed that she had only published the work after some coercion, and then only under the cover of anonymity (although her cover was soon blown). Such declarations were common for both men and women, as writers worked to protect themselves from attack by appearing modest.

Modesty was not the only arrow in Du Châtelet's quiver: she also used her maternity to deflect criticism of her work and herself. In the preface of the book and in her correspondence, Du Châtelet stressed that she had written the *Institutions* because maternal duty demanded it. Her son required a textbook on natural philosophy, but no such book had existed until she wrote the *Institutions*. The marquise's self-presentation did more than establish that she could be both a mother and a thinker; she argued that the two were mutually beneficial. The preface made clear that her work on the *Institutions* proffered direct benefits for her son's well-being: "I will think my labors well directed if they can inspire in you [my son] a love of the sciences and a desire to cultivate your reason. What pains and concerns parents will undertake in the hope of procuring honor for and increasing the fortunes of their children!" In the very next sentence, however, Du Châtelet insisted that her work on the *Insitutions* was for *her* intellectual benefit as well as her son's: "Are not an understanding of truth and the habit of researching it worthy of my time, especially in an age when a taste for natural philosophy permeates all ranks . . . ?"[43] Writing the *Institutions de Physique* furthered Du Châtelet's ambitions for her son but also for herself, and in a way that guarded her from insult. In crafting her public persona, Du Châtelet charted the same course as male senti-mental savants: she stressed that she was admirable because she was both engaged and sensitive. Her private virtue testified to her suitability for a public role.

She continued to emphasize her maternity when she learned that the authors of the *Portrait Gallery of Contemporary Authors Famous for Their Learning*, Johann Jakob Haid and Johann Jakob Brucker, wished to include her in their volume. Du Châtelet was flattered by the invitation. This text, published in Augsburg between 1741 and 1755, included engraved portraits and biographies of the one hundred individuals chosen for inclusion. To be recognized in so public a fashion for her learning appealed greatly to Du Châtelet. Her letters on the subject ring with enthusiasm: here was a wel-come change of pace from mockery and condescension.

Du Châtelet wanted to make sure that her portrait presented her in the most flattering possible light. In particular, she wanted to play up her maternity. This was a somewhat unusual request for the authors of the *Portrait Gallery*. Owing to space constraints, their entries generally did not discuss personal details related to marriage and family life. But in one of several letters she wrote to Johann Bernoulli (acting as intermediary for Haid and Brucker), Du Châtelet suggested that they "add that I have only two children, that last year I married my daughter to the Duc de Monte-nero of the house Caraffa, and that I wrote the *Institutions phisiques* [sic] for my son's education; he is fifteen years old and a musketeer."[44] They might also note, she suggested, that she corresponded with a panoply of well-established savants. Despite the fact the authors usually avoided such details, Brucker heeded Du Châtelet's request. He must have shared her sense that highlighting her maternity would give her the best chance of earning a positive reception.[45] Social connections plus private virtue equaled a successful self-fashioning strategy.

Du Châtelet had long struggled to reconcile her work with her femininity. Her prior efforts to intervene in scientific debates had been mocked as silly and pretentious. To cope, the marquise experimented with different means of representation, including anonymity and aristocratic virility.[46] In this instance, she couched the *Institutions* within the domestic realm and made plain the maternal ambitions driving her work, which made the text more compatible with early modern gender norms.[47] To prevent charges that the work was unserious, Du Châtelet chose not to inflect the text with levity or embellishment. She insisted that she had drawn on her "raison" rather than her "esprit" in composing the book; this was not a silly project. The *Institutions* thus had a complicated relationship with Du Châtelet's femininity: she wrote it as a serious work of natural philosophy, a genre many assumed to be the exclusive purview of men, and yet she overtly highlighted her maternity and love for her son. Confident that this form of self-representation gave her some cover, Du Châtelet pulled no punches in writing the text, taking on authorities such as Dortous de Mairan (secretary of the Academy of Sciences). Although some persisted in lampooning the book as derivative, others saw the *Institutions* as a major accomplishment and respected Du Châtelet's intelligence and innovations.

In addition to her son, Du Châtelet also had a daughter, Gabrielle-Pauline, born in 1726. Du Châtelet refrained from writing about her daughter's upbringing in great detail.[48] She lived in a convent, like many girls (in this, Du Châtelet parted ways with other philosophes). Convents functioned as religious schools and safe havens for virgins. The young girl also visited her home, where she studied Latin and Italian and acted in Voltaire's plays during small family productions. That polish made it possible for Gabrielle-Pauline to pursue traditional avenues of social advancement: exchanging letters with learned and highborn correspondents, acquiring a position as a lady in waiting, and securing an aristocratic and wealthy husband. She married the Duc de Montenero-Caraffa, a Neapolitan aristocrat, when she was seventeen years old. This marriage, arranged by her parents, bestowed financial, social, and political power on the young woman.[49]

Du Châtelet gave every appearance of being pleased with her daughter's upbringing, but readers familiar with her translation of Mandeville's *Fable of the Bees* might feel surprised by the traditional and limited nature of Gabrielle-Pauline's education. In her 1735 preface to the work (written when her daughter was eight years old), the marquise argued that education, not nature, determined intellectual differences between the sexes. Men did not enjoy more natural intelligence nor were they inherently more rational. They simply had better schooling. If girls' education would become more rigorous, then girls would become more rational.

Despite this passionate plea that girls should benefit from the same education that boys received, the Du Châtelet children received very different educations. Why? Du Châtelet's pedagogical choices were rooted in a broader cultural framework: she wanted to make her children useful and honorable, and she gave them the educations that would best accomplish those goals. Noble merit and enlightened utility intersected here. Florent-Louis would best make himself meritorious and useful if he studied physics and prepared for a military career, and Gabrielle-Pauline would best shine if she made herself into a polished spouse.

Du Châtelet's education of her children was her entrée into ongoing debates about nobility and civic virtue as well as an opportunity for her to cast a more flattering light on herself and her work. In this, she was like many of the male sentimental savants discussed in this book. But Du Châtelet's path was an especially treacherous one because she was a woman. She needed to stress her domestic bona fides to a greater extent than did male savants because learned women were more likely to weather accusations that they were unnaturally ambitious and unfeeling. Women could be virtuous, of course; in the second half of the century, they were even thought to be more naturally virtuous than men.[50] But that natural virtue was entirely demonstrated by domestic care and affection; *femmes savantes* seemed to subvert this natural goodness by trying to think and act like men. By emphasizing that she did love her children and did devote herself to their education, especially in her son's case, Du Châtelet hoped to forestall some of these criticisms. She was no monster—she was a learned and engaged noblewoman motivated by her search for scientific truths and her love for her children.

## DENIS DIDEROT

Du Châtelet displayed few doubts about how to raise her children but Denis Diderot was more ambivalent. How far should he push his daughter Angélique's education? He eventually settled on marriage and motherhood as the ultimate focus of her education, but he adopted unconventional methods to prepare her for those roles. Angélique, the only one of

Diderot's three children to survive past infancy, was born in 1754. Her early education was a source of contention for Diderot and his wife, Antoinette Champion. The couple had married for love, but once that love faded, they could only focus on the other's faults. By the time Angélique was born, their home was divided by religious disagreements. At first, Madame Diderot assumed control of Angélique's education. Although typical, this arrangement troubled Denis. He saw his wife as irrationally devout, bad-tempered, and unsupportive of his career; Antoinette in turn despised his flippant attitude toward religious faith. It seemed to Denis that his daughter was far removed from him, stuck in her mother's world and bogged down in exercises of religious devotion.[51] Diderot, a philosophe who distrusted organized religion, found this troubling. "I am so charmed by my little girl, because she reasons through everything she does," he confided in a friend. "[But] what a shame that her education corresponds so poorly to her natural talents!"[52]

The problem, he suggested, was his wife's guidance. "What a wonderful woman [Angélique] could be one day! But all that awaits her from morning to evening are gibes and silly remarks. Whatever I might do with her one day, that first bad impression [*incrustation mauvaise*] will always remain."[53] One day, in a foul mood, he sniped: "Truthfully, if the child were carried away by some violent illness, I am not sure if I would regret it. It would be better that she died than be abandoned to the mercy of her mother."[54] Like many, Diderot believed that the experiences of early childhood left indelible marks, and he did not like what he saw.

To salvage what he could, he demanded greater pedagogical control in 1762, when Angélique was nine years old. Once at the helm, he felt confident that he had the power to mold his daughter as he saw fit. He marveled at her superior mind and dotted his correspondence with exclamations about her. "I have a sensible and rational child"; "What an imagination she has! How reflective she is! How wonderfully she sees the world!" Her good heart complemented her quick mind: like her sensitive father, Angélique was easily moved to tears.[55] She appeared frequently in her father's correspondence, and his letters represent their relationship as tender and loving. As mentioned in chapter 1, Angélique was a key part of her father's sentimental image. He enjoyed describing himself as a loving *père de famille* and found it a useful way to convey his virtue and sensitivity.

Once Diderot had taken over his daughter's education, he had to decide what she should learn. He featured this dilemma in *Rameau's Nephew*, a fictional dialogue between "Moi" and "Lui" (the titular nephew of the musician Rameau). The pedagogical choices up for debate fell into two categories: encouraging Angélique to be "pretty, amusing, and attractive" or to raise her "to reason correctly," which "Moi" judged "an uncommon thing in men and even rarer in women." The study of "grammar, literature, history, geography, a bit of drawing, and a great deal of ethics" would accomplish this goal.[56] The debate in *Rameau's Nephew* suggests some hesitation when it came to education: How far should Diderot push the limits placed on girls' education?

This hesitation was most apparent in his discussion of reading material. In one letter, he proudly noted, "I entrusted her with a difficult work for her age [which she understood and reasoned with]."[57] She "devoured" all sorts of books, which Diderot enjoyed discussing with her over long walks.[58] On another occasion, he mentioned "a sublime treatise on human nature, a treatise which I would recommend to my child, to my friend" written by Thomas Hobbes. This treatise, though apparently "the thorniest, most difficult, the most argumentative, the most abstract material," was still "very useful for my daughter and my friend."[59] That he recommended such a text to his daughter suggests how confident he was in Angélique's ability to read and think at a high level. His choice of reading material also testifies to his willingness to expose her to unconventional selections.

Seeing how readily his daughter grasped difficult texts, Diderot began to dream of forming her into a miniphilosophe. It would be all too easy, he mused, to introduce her to the audacious ideas he favored: "What a trail I could blaze with a mind like hers, if I dared. It would take nothing more than leaving a few books lying around."[60] A gleam in his eye, his imagination was alight with thoughts of showing off his daughter to his friends.

Yet he ultimately dared not. Wary of making his daughter too radical a thinker and nervous about sexual mores in particular, Diderot kept a close eye on Angélique.[61] His own fondness for lurid materials—this was, after all, the writer who gave talking vaginas a starring role in one of his first publications—dissipated when it came to his child's upbringing. When his friend J. B. A. Suard sent salacious materials to Diderot's home, Denis fumed: "I tried to make him understand the possible consequences of his actions: the corruption of my daughter and my eternal hatred."[62] And in the same letter in which he had exclaimed "What a trail I could blaze with a mind like hers, if I dared," Diderot recounted a conversation in which he had conveyed to his fifteen-year-old daughter the absolute necessity of guarding her sexual virtue. Gallantries, he warned her, "are as if the man meant to say . . . *Mademoiselle, if you would be so kind, out of consideration for me, to please dishonor yourself, lose all social standing, banish yourself from the world, lock yourself in a convent for the rest of your life, and make your mother and father die of sadness.*"[63] Diderot's dreams of trailblazing stopped well short of encouraging sexual license. His fear of ridicule guided him toward marriage and motherhood as the best choice for his little girl.

His efforts to protect his daughter from illicit sexual materials and to exhort her to remain a virgin until marriage are hardly notable in and of themselves. Virginity was central to ideas of female virtue and sexual honor. What is noteworthy is that Diderot completely rejected the standard means to ensure virginity: convent education. Diderot ranks as one of the most anticlerical thinkers of the Enlightenment, and he found much to criticize in the Catholic Church. He disapproved of celibacy, a practice he considered socially useless and selfish, and he believed religious seclusion ran counter to human nature and harmed those forced to endure it. He especially loathed convents, in part because his beloved sister had died while living in one.

Diderot expressed his disdain for religious life in his novel *The Nun*.[64] The story revolved around the trials of Suzanne Simonin, a young woman compelled to take religious orders. Suzanne endured a variety of hardships, including her mother superior's attempts at seduction. Suzanne, who lacked any concept of sexual intercourse, could not understand what was happening to her. When the mother superior forced Suzanne to caress her and eventually achieved orgasm, Suzanne was baffled and could only assume that the older woman had fallen ill. Diderot used this episode—depicted as "unnatural" because it involved two women—to demonstrate the dangers of convent life. "There you have the effect of segregation," he thundered. "Man is born for life in society; separate him, isolate him, and . . . his character will go sour, a hundred ridiculous affections will spring up in his heart, [and] extravagant notions will take root in his mind."[65] Convent education was clearly out of the question for Angélique.

What would fill the void left by convent education? Diderot decided anatomy lessons would serve nicely. This was an ostentatious move: he rejected religious education and replaced it with a scientific one. He reasoned that his brand of education would better cultivate Angélique's virtue and make her into the kind of sensitive, knowledgeable woman that society needed. Studying the mechanics of the human body would guard Angélique against misinformation peddled by gallants and rakes and would prepare her for her eventual marriage and pregnancies. And Diderot knew just the teacher: Mademoiselle Biheron, a noted anatomist who crafted wax models famed throughout Europe for their astonishing lifelikeness. In 1771, at the age of seventeen, Angélique received her first "leçon d'anatomie," which she "completed with ease and without distaste."[66]

Why anatomy? The growing popularity of dissection probably influenced Diderot's decision. In Italian cities like Bologna, crowds would gather for public anatomy lectures, straining to see a cadaver laid open on a marble table, listening to the Latin disputations of university professors, and breaking into riotous laughter or taunting. Anatomy fascinated the public, who delighted in inspecting the innards of the body before them. And anatomy did more than thrill. It seemed an especially useful branch of science, with the potential to better train medical professionals and advance scientific understandings of the human body.[67] A major problem with anatomy, however, was that corpses decayed. The lessons they taught were fleeting. Hence the interest in wax models, like those carved by Mademoiselle Biheron or her counterparts.[68] Thanks to these wax models, students could reap the benefits of anatomical education without having to suffer the smell and feel of decaying human flesh.

Diderot shared this enthusiasm for anatomy, and he even found new ways to make the study of anatomy useful. By learning medical facts rather than remaining unaware of how sexual intercourse worked, Angélique's curiosity would be nipped in the bud and she would not be tempted to seek out sexual experiences.[69] Unlike doctors—whose curiosity was piqued by cataloguing the parts of the human "machine" and who were spurred to discover new ways to treat disease and perform surgery—Angélique would be made less curious and less likely to test the waters of premarital sex.[70] Eliminating her ignorance in no way compromised her innocence. Moreover, she would enter her marriage equipped with the knowledge of how sex and bodies functioned and would not suffer from feelings of fear or shame.

What might these lessons have been like? Diderot described a "company" of his friends who sent their children to take the lessons together.[71] Perhaps students crowded around Biheron and her anatomical models in much the same way that spectators gazed on the bird in an air pump in Joseph Wright's 1768 painting (figure 4.1), with a small crowd gathered around the anatomist—curious to see more, dazzled by the verisimilitude of the objects, and astounded by what they saw. In Diderot's telling, however, young girls did not turn away from the spectacle but observed it unafraid. Anatomy was both useful and captivating.

Angélique's anatomy lessons were soon implicated in the ongoing feud between Denis and his brother Didier, a priest who had long harangued his sibling for his unorthodox views. Angélique's education provided a key opportunity for Denis to represent himself and his daughter as paragons

Fig. 4.1. Joseph Wright of Derby, *An Experiment on a Bird in the Air Pump*. 1768. Image copyright © The National Gallery.

of virtue, contra his brother's assumptions. Unlike his celibate brother, Diderot had married and fathered a child; he had then raised that child to be a moral and lovable woman. Didier may have thought his brother was dangerously irreligious, but Denis recast himself as the more virtuous and useful of the two siblings. "M. l'abbé," Denis declared, "that which one learns in a convent, [my daughter] knows better than those who have passed their whole youth in one, perhaps even better than you yourself." Even better, "she is not stupid or pretentious like those girls. She has maintained—and I hope she will for her whole life—the simplicity, the sweetness, the innocence of a young girl." How, Didier might wonder, had Denis formed his daughter into such a woman? By organizing "three anatomy courses that she completed at the home of one Mademoiselle Biheron."[72] Anatomical education had made his daughter more virtuous and more innocent than a convent education ever could have. Angélique's education provided Diderot with a major opportunity for self-fashioning.

The brothers never reconciled, but Diderot also lobbied more sympathetic friends and associates. He told John Wilkes that Biheron was a masterful craftswoman and urged him to enroll his daughter in similar lessons.[73] He likewise pleaded his case before Catherine II, who was at this time Diderot's patron and who had recently founded a system of schooling for noble girls. Diderot praised her efforts, effusing that Russian women would now receive educations superior even to the one he had been able to give his own daughter. With one exception, that is: the Russian curriculum did not include anatomy courses.[74] Diderot again mentioned how successful Angélique's lessons had been, even relating a story in which Angélique had discovered naughty reading materials at home but had shrugged off the more salacious elements. Her anatomical education had inoculated her against corruption.[75] Diderot encouraged Catherine to consider inviting Biheron to her court. He also contacted one of Catherine's ministers, General Betzki, and conveyed Biheron's own ideas for anatomy classes. Catherine's schools would reach many more young women than Diderot could on his own, and might serve as a model for other enlightened monarchs.

Anatomy lessons thus allowed Diderot to pursue the traditional goal of preserving his daughter's sexual virtue in a way that accommodated his most cherished ideas. Providing Angélique with an anatomical education in place of convent education was secular rather than religious, social rather than secluded, and scientific rather than dogmatic. He then used his personal experience to lobby his friends and patrons to follow his lead. They, too, could be paragons of sentiment, secular virtue, and public utility.

Her anatomy lessons complete, Denis deemed Angélique ready for marriage. Here, too, Diderot found himself on familiar philosophical ground. He wrote often of marriage, with a corpus that ranged from a total rejection of matrimony to sentimental celebrations of loving marriages.[76] In works such as Le Père de famille, Diderot insisted that children should decide when and whom to marry and that they should base this decision on a deep emotional attachment. At first glance, however, Diderot behaved like a traditional patriarch in finding a match for his daughter. He allocated himself a principal role in this search and even tried to arrange marriages with his fellow philosophes (all much older than Angélique). Diderot eventually abandoned the idea of a philosophe-husband and settled on Caroillon de Vandeul, a reasonably prosperous young man from Diderot's hometown of Langres.

Yet although Diderot accorded himself significant authority in the process, he also solicited Angélique's input. He defined himself as his "daughter's master," but also saw himself as limited by his daughter's wishes. Father, mother, and daughter were mutually constrained: "we must be in

agreement."[77] The choice of Angélique's husband was a collective decision.[78] Bound by love and duty, no member of the family would disregard
the wishes of any other. Although his plays tended to contrast parental
dictates with children's desires, Diderot imagined a more harmonious process for his own family. He expected the oft-competing interests of love
and money to fit together easily in his daughter's case. He did not want
Angélique to choose between a practical relationship and a loving union (a
choice often foisted on the protagonists of eighteenth-century literature);
like many other parents, he wanted her love match to also be a practical
choice.[79] Nevertheless, Diderot remained uncomfortable with the dowry
process. He felt outraged when his future son-in-law had the temerity
to negotiate terms, and fulminated that "Angélique does not like to be
bought and sold. She loves Caroillon; but if he continues [to dwell on the
dowry negotiations] she will scorn him, and goodbye to love. I wish I could
stop her from marrying him."[80]

That feeling of regret only intensified. Diderot had high hopes that his
daughter's marriage would be enlightened and companionate. Two young
people, united by common affection and reason, would spend their days
quietly enjoying each other's company and reading books suited to their
own particular interests. Diderot hoped that Caroillon would appreciate
Angélique's musical gifts and encourage her to devote more time to hon-
ing her skills.

The Vandeul marriage did not live up to this fantasy. Diderot missed
Angélique's company terribly. His daughter's marriage did not bring him a
sense of relief and pride but instead plagued him with feelings of loneliness
and abandonment. The thought that he had been replaced in his daughter's
heart disconcerted him greatly. Worse still was his suspicion that he had
been supplanted by an unworthy individual. His son-in-law proved a disappointment because he neither nurtured Angélique's musical talents nor
encouraged her to read widely. His only interest seemed to be dressing his
wife in the latest fashions and showing her off to his friends. Diderot wrote
a disconsolate letter to Grimm, confiding his grief in seeing his child live
such a confined life: "I introduced my child to reflection, to reading, to
the pleasures of the quiet life, and taught her to scorn . . . frivolities. . . .
But this little man dresses her up like a doll, and she has to spend her day
decorating herself so that she can keep him happy." He doubted that his
son-in-law's intelligence came close to his daughter's, and he grieved that
such a witless clod controlled his child: "This shows me that my daughter,
who has a mature mind, is being contradicted by a man with the mind of
schoolboy. . . . My friend, he is working to turn my child into a dull, imper-

tinent little fool who thinks of nothing more than how to place a *pompon*, simper, gossip, and smile. This grieves me."[81] Diderot felt frustrated by his daughter's marriage and outraged that he could do nothing.

Diderot's frustration stemmed in part from his understanding of feminine virtue, a concept that sparked debate during the Enlightenment. In the first half of the eighteenth century, the *honnête femme* was the model of virtuous womanhood. Respectable and restrained, adept at navigating the treacherous world of the court, she kept her true feelings hidden. After the 1750s, however, sincerity ruled. The ideal woman had no guile and was easily read by all who knew her. Her transparency revealed her to be uncorrupted by society, her natural virtue still intact. This was not a woman likely to be swayed by the frippery of high society; luxury held little appeal for her. She was a creature of natural *sensibilité*, immune to the affectations of *le monde*.[82] When Caroillon encouraged Angélique to ignore the delights of quiet reading and to instead seek pleasure in elaborate outfits, he undermined the natural virtue and sensibility that Diderot had cultivated in her. From her father's perspective, Caroillon was wooing Angélique to the dark side.

His distress fueled a bitter composition on women and their place in society, titled "Sur les femmes" ("On Women," 1772) and published in the *Correspondance Littéraire*, in which he ruminated on the limitations imposed by marriage and motherhood. He implicitly linked this essay to his daughter by repackaging the conversation in which he counseled Angélique to zealously guard her sexual virtue. The original version had claimed that when a gallant spoke of love, what "the man meant to say [is:] *Mademoiselle, if you would be so kind, out of consideration for me, to please dishonor yourself, lose all social standing, banish yourself from the world, lock yourself in a convent for the rest of your life, and make your mother and father die of sadness*."[83] The text of "Sur les femmes" is strikingly similar to that discussion:

> But what does the phrase *I love you*—so easily stated, so frivolously interpreted—really mean? It actually signifies: "If you would please sacrifice your innocence and your virtue to me; lose the respect of others and for yourself; walk with lowered eyes through society, at least until the habit of libertinage gives you the effrontery to do otherwise; renounce any honest station in society, [and] make your parents die of sadness just so that you can grant me one moment of pleasure, I would be truly obliged."[84]

Angélique was on Diderot's mind. His love for his daughter and anger over her marriage drove him to lament woman's lot in life: "What, then, is a woman? Neglected by her spouse, ignored by her children, with no role in society . . . women are oppressed by the cruelty of civil laws, combined with the cruelty of nature; they are treated as if they were stupid children."[85] Women were not inherently stupid or frivolous, but the men controlling them often made them that way. And in such a patriarchal society, what chance did women have to escape this pernicious influence?

Diderot did not mention Angélique by name in "Sur les femmes." But the structural similarity between this text and his earlier advice for Angélique suggests that she was there in spirit. From Diderot's perspective, his education of Angélique had gone awry despite his best efforts. His disappointment led him to doubt that women could ever rise above their subordinate state: How could they, with so many unenlightened fathers and husbands exercising control over them?

Angélique Diderot's education shows her father torn over the choices available to him and to her. He wanted his daughter, his only child, to grow into an intelligent adult who read widely and challenged her mind. He wanted for her to be a useful citizen, and for her virtue to reflect well on him, her loving father. But he was, at the same time, allergic to accusations of impropriety. He did not want his daughter ridiculed, nor did he wish to suffer criticism. Instead of urging her to become a *femme philosophe*, he instead pushed her to become a paragon of female virtue: an enlightened wife and mother. Anatomy lessons constituted an innovative way to accomplish this goal, but the results of Angélique's carefully considered upbringing were mixed. Diderot was proud of his beloved daughter, but devastated by her marriage and the intellectual and moral limitations it seemed to place on her.

The ups and downs of Angélique's upbringing made their mark on Diderot's published writing. They lent empirical weight to his educational ideas, his attacks on convents, and his advocacy for anatomy lessons. They also gave Diderot another opportunity to represent himself as a virtuous philosophe: his love for Angélique testified to his *sensibilité*; his devotion to her education, and to education in general, demonstrated his public engagement. But Diderot's experiences as a father also inflected his writing with despair that women would remain morally and intellectually stunted as a result of their stultifying marriages. He anguished over the problems women faced, but could not find a way out.

## LOUISE D'EPINAY

Louise d'Epinay's education of her daughter and granddaughter fits within a similar framework of civic virtue and public utility. D'Epinay, born in 1726 to the governor of the citadel in Valenciennes, married Denis Joseph de la Live d'Epinay in 1745. They had three children, two of whom survived past infancy: Louis-Joseph and Angélique-Louise-Charlotte. The marriage quickly soured, spoiled by her husband's extramarital affairs and profligate spending. In 1749, D'Epinay secured a *séparation de biens* from her husband, severing her finances from his.[86]

Her marriage in ruins and her health a constant problem, Louise d'Epinay sought consolation in her studies. Her circle of friends expanded to include such luminaries as Duclos, Diderot, Grimm (her lover), and Rousseau, at least until they had a famous falling-out.[87] She published three pedagogical texts: *Letters to My Son*, "Letter to My Daughter's Governess," and *Conversations with Emilie*. She found the ideas of a nature-based education compelling, and clear links existed between her works and those of Rousseau. Unlike Rousseau, however, D'Epinay insisted that mothers should guide their children's educations past their infancy and argued that girls' education required a moral and intellectual overhaul. Women would shape the future of the nation through their husbands and children, and they needed an education to prepare them for such a task.[88]

She first published "Letter to My Daughter's Governess" in 1756, when her daughter Angélique was seven years old.[89] The letter describes the regimented schedule and "general rules" that D'Epinay planned for the young girl and her governess. Angélique should begin her day with prayer and catechism, and then head outdoors to walk in nature. There, the governess should encourage her charge to enjoy the beauty of nature and to let her curiosity guide her, perhaps to watch insects at work and learn a bit about the virtues of industriousness. They would then engage in constant conversation, with the governess taking cues from the child's own interests and questions. This method would teach Angélique more than "all the masters put together." If the child said something wrong, the governess should go beyond mere correction by engaging Angélique in rational debate, presenting evidence and logic to support her cause. Punishment or reprimands were discouraged. In the afternoon, the young girl would study history and geography, embroider, and visit female relatives. By carefully cultivating her mind and heart through conversation, love, and reason, Angélique would develop into an intelligent and moral woman. This would prepare her for her ultimate destiny as a wife and mother: "one day, she and her husband will thank us."[90]

D'Epinay also raised a son, Louis-Joseph, and discussed his education in her *Letters to My Son*, published in 1759. This work consisted of twelve letters addressed to her son. D'Epinay strongly advised her son to live a virtuous life, defined by her as one of sincerity, industriousness, and compassion. She feared that her son, who already exhibited a troubling fondness for luxury and disdain for hard work, was too much like his pleasure-loving father. Each letter expounded on some facet of moral development: "on lying," "on nature," "on compassion," *et cetera.* She extolled the advantages of a rural lifestyle (founded on an appreciation of nature and its simple pleasures) as opposed to the evils of urban society (based in silly frivolities and costly luxuries). Unlike "Letter to My Daughter's Governess," these letters were not addressed to a tutor; D'Epinay wrote directly to her son and thus positioned herself—rather than a male family member or a religious figure—as his moral compass.[91] Whereas Rousseau worried that women lacked the willpower to discipline and instruct their sons, D'Epinay placed mothers at the center of pedagogical reform. She believed that she, more than anyone else, was qualified to guide her son's moral development.

Despite D'Epinay's best efforts, Louis-Joseph turned out to be a rake and a spendthrift—in short, a total disappointment. His family even demanded his imprisonment by means of a *lettre de cachet* (the Old Regime policy of allowing families to lock up their subordinates).[92] D'Epinay took her son's failures badly: "I have lost all heart. Most of the time, I reflect sadly on what the future will bring."[93]

Distraught, D'Epinay reached out to her close friend the Abbé Galiani. Unfortunately, he included a heavy dose of sarcasm with his consolation. Louis had inherited these bad traits from his father and there was nothing D'Epinay could have done about that. "What in the devil made you have children with M. d'Epinay! . . . Are you so delirious as to believe in Rousseau and his *Emile*, and to think that education, maxims, and speeches would do anything to change how his head is organized?"[94] Galiani then snapped that if education was really so all-powerful, perhaps she should go catch a wolf and raise it to be a dog. The Italian found French readers all too willing to buy into the fashionable concept of human perfectibility.

Yet while D'Epinay's sensationalist beliefs must have been shaken by this terrible stumble, she did not adopt Galiani's stance that nature mattered more than nurture. Instead, she decided that the fault lay in insufficient discipline and diligence on her part. In 1770, when it had become abundantly clear that her son was not the man she hoped he would be, she wrote that "education only succeeds when I have forced it to conquer difficulties by careful application and assiduity."[95]

In her last and most popular publication, the *Conversations with Emilie*, D'Epinay depicted herself (one of the protagonists in the novel) as an ever-watchful and morally austere figure; *this* child, at least, was not going to follow Louis-Joseph on his path of moral dissipation. Unlike her previous works, in which she dictated her recommendations, D'Epinay here experimented with a new format: dialogue. In the *Conversations*, the active, inquisitive Emilie walked, ran, and played, all while questioning everything. The mother kept on top of her every move, fully engaged in educating her daughter properly. This total devotion inspired D'Epinay's editor to describe education as an exercise both "sweet and aggravating [*pénible*]" for the mother.[96]

Emilie from *Conversations with Emilie* was not fictional, unlike Rousseau's Emile. D'Epinay made much hay out of this distinction. She stressed the verisimilitude of her works and insisted that all letters and conversations were lifted directly from her life.[97] Her contemporaries admired this component of her work. Her editor praised her for modeling "these Conversations after those that took place between mother and daughter"; the Abbé Galiani applauded her work as a "true production of the heart."[98] The second edition even won the Montyon Prize, awarded by the Académie Française for the "best written and most useful" book published that year.[99] D'Epinay depicted her philosophical ideas as emanating from her maternal duties, and her readers responded with enthusiasm. The intellectual trends of the time favored empirical work suffused with sentiment, and it was all for the better if the author was virtuous and had firsthand experience with the subject. She used her maternity as a sword to cut down rival authors, especially those like Rousseau who lacked practical knowledge.

Conscious of her audience, D'Epinay almost certainly exaggerated the reality of her work. Her daughter, Angélique, wrote to Galiani that "*Maman* diminishes her glory when she says that my daughter made all those comments. It is very true that she said some of them; but if *Maman* had not had it in her head that she should arrange, trim, and expand those comments, the work would have been quite bad."[100] D'Epinay also obscured her real-life relationship with Emilie. Emilie was her granddaughter, but in the *Conversations*, D'Epinay styled herself as Emilie's "Maman." Centering the book around a mother-daughter relationship made it more relatable and made D'Epinay the central character in the young girl's life.

D'Epinay's exaggerated verisimilitude emphasizes how important it was that her text appeared to be an accurate chronicle of her family life.

Furthermore, her focus on the particular versus the abstract, the adaptable versus the rigid, not to mention emotional attachment rather than clinical detachment, places her within the tradition of sentimental empiricism.[101] Indeed, D'Epinay stressed that her work was "neither a general rule nor a system."[102] As Jessica Riskin has shown, savants working in the age of sentiment often railed against "systems," abstract schemes that bore no relationship to particular facts. D'Epinay's language is thus telling. She implied that her ideas were superior to those of more general theorists, who had not tested their ideas on actual children. Her editor agreed, as he demeaned the utility of general theories of education that were "of little resource" and "did not indicate any precise route" for mothers to follow. D'Epinay's conversations were not intended as a script for other mothers to follow but rather modeled how parents should engage with their children.[103]

Her authority established, D'Epinay presented a series of conversations in which she ("Maman") instructed Emilie to eschew frivolity, luxury, and pride and to look instead to nature for guidance. Previously she had issued bald declarations, as when she expounded to her son that "a sensible man knows to monitor his amusements and I would be truly anguished if the taste you show for music would ever cause you to neglect more serious occupations."[104] The *Conversations*, however, relied more overtly on examples drawn from nature to convey her points. For example, at one point Emilie caught a fly and wanted to keep it as her pet:

EMILIE: I'm going to remove its wings so it can't fly away, and then I will take care of it.

MÈRE: Wait, my dear friend. Has it bitten you? Has it hurt you?

EMILIE: No, Maman.

MÈRE: And so why would you want to hurt it?

EMILIE: But it wouldn't hurt.

MÈRE: To rip off the fly's wing would be as if someone cut off your foot or your hand. You think that because you cannot hear its cries that it does not suffer? You are mistaken. That is a sensitive creature, just like you, and you have no right to hurt it.

The mother pressed the issue further, seizing the opportunity to discuss more abstract concepts: "Always remember, Emilie, that one must not rely on force except to protect the weak, not to oppress them. . . . You must respect the sensibility of even the smallest productions of nature. A fly, a june bug, a dog, a tree, all are nature's works."[105] The simple act of Emilie catching a fly evolved into a conversation about compassion, cruelty to animals, and the just and unjust uses of force. Likewise, Emilie's delight at seeing monkeys and D'Epinay's distaste for a species she labeled "lying, treacherous, malignant, thieving" creatures morphed into a discussion of human beings as reasoning animals.[106] Nature served as a useful class-room in the *Conversations*. It was a site for productive interactions be-tween mother, child, and creation.

D'Epinay was bluntly honest in these conversations. At one point, she told Emilie that she was not pretty as a way to teach her about the dangers of flattery and false praise.[107] In another conversation, she told Emilie the story of a boy who had stood on a chair, despite his mother's command that he never do so, in order to sneak some jam. He slipped and fell but, anxious to avoid punishment, did not tell his mother. He developed head-aches and fever and died two days later. Honesty was a matter of life and death in the *Conversations*.

These educations were meant to safeguard children's natural virtue. She did not exhort them to classical heroics; she did not impress upon them the necessity of conforming to social expectations. On the contrary, her writings show a deep suspicion of society and the corrupting influence that it exercised over children. She did not wish to see her daughter and granddaughter preen before their admirers. She did not want her son to fall for the charms of high society instead of the simple pleasures of rural life. The unsubtle moral message conveyed in all her pedagogical writings fo-cused on cultivating sincerity, transparency, and civic virtue. These were educations strongly marked by eighteenth-century ideas of natural virtue and sensibility.

Like a true heir of Locke, D'Epinay encouraged her children to develop their minds and bodies by exploring outdoors. Like Du Châtelet and Di-derot, D'Epinay emphasized the value of a nature-based and utilitarian education, although the subjects she chose for her children were more conventional than natural philosophy or anatomy. Again like her fellow philosophes, D'Epinay understood her pedagogical program as preparing her girls for marriage and motherhood. Despite her personal difficulties with matrimony, she could not imagine her daughter or granddaughter not marrying; all that she could do was hope for the best.

D'Epinay's self-fashioning is very revealing of late eighteenth-century intellectual culture. She claimed that every letter, every conversation was drawn from real life (although we should take that assertion with a grain of salt). With even greater enthusiasm than her peers, many of whom also relied on a rhetoric of personal experience, D'Epinay tore down the wall separating public from private. She encouraged readers to take her "real life" experiences as a model for their own behavior. In insisting on the true-to-life nature of her work, D'Epinay was influenced by multiple pressures. Like Du Châtelet, she found that emphasizing her domesticity made her authorship more acceptable, so she also joined the tradition of women writing as wives, mothers, and daughters. But as we have seen, this rhetorical posture was not limited to women in the eighteenth century. Men of letters also claimed that their family lives demonstrated their virtue and therefore their suitability for public life. This understanding of virtue as founded in sensibility and private life also influenced D'Epinay. Both of these traditions shaped D'Epinay's texts and together they explain the extraordinary emphasis that she placed on personal experience.

## THE MARQUIS DE CONDORCET

Condorcet's hopes for his daughter were marked by his feminism. In his writings and speeches, he argued that perceived gender differences in intelligence or sensitivity—men allegedly having more of the former, and women the latter—sprang from education and environment. In this, he parted ways with most male revolutionaries. He even advocated women's political equality and right to vote.[108] Moreover, he urged the National Assembly to provide a basic primary education without charge to boys and girls alike, because only public coeducation would result in universal liberty and equality.[109] These schools would endow future citizens with practical skills like reading, writing, and calculating sums and provide a foundation in civics, agriculture, and various trades.[110] Condorcet's schema remained hierarchical—he imagined that some pupils would be more talented than others, and that the state should invest more in their educations—but it was egalitarian in his insistence that all students have access to education and that intelligence could be found in any social milieu.[111]

Many revolutionaries beyond Condorcet put their own spin on Lockean sensationalism. They were not interested in gradual reform or preserving existing institutions; they wanted to bring about a fundamental shift in the French and, indeed, all humanity. Pedagogy became one means for them to bring about a major political transformation. The state could help transform the French from subjects into citizens by working through culture: festivals and monuments would provide citizens with models to follow, classically inspired clothing would encourage citizens to practice classical virtue, and Revolution-themed crockery would serve as an everyday reminder of political principles. These practices all assumed that human beings were fundamentally malleable, that their most intimately held beliefs—indeed, their very natures—could change with careful guidance. And as children seemed especially malleable, having never known the corruptions of the Old Regime, education offered a transformative way to shape the French nation.

Condorcet did not implement his plan. A member of the National Assembly, he denounced Jacobin policies and, as a result, faced arrest, trial, and execution. He evaded authorities and hid in Paris for several months. Although his education plans came to naught, he spent his last months contemplating one education in particular: that of his young daughter, Eliza. Condorcet's plans for his daughter illuminate his goals for her and presumably other girls.

Condorcet urged Eliza to remain an independent woman, further separating himself from the crowd of men writing about women in the age of the French Revolution. Most men assumed that women were incapable of independence. Headstrong women would at best flounder and at worst sow social disorder. Male revolutionaries preferred to think of women as attached to men: fathers, husbands, and brothers. But that was not the life Condorcet wanted for his little girl. He pleaded with Eliza to maintain a degree of independence, to acquire "the habit of work, so that you can be self-sufficient, so that your work can provide for your needs, so that even if you are reduced to poverty, you will not be reduced to a state of dependence." Even if Eliza should find herself in a fortunate position of financial security, she should still earn her own living. "This resource . . . will serve you well by guarding you from fear, sustaining your courage, allowing you to look with an unblinking eye on any reverse of fortune you might have to endure."[112] Eliza should not depend on her husband or anyone else to provide for her. She should provide for herself and set her own path. Condorcet's letter reflected his belief that women should aspire to be active, independent citizens. For him to insist that his daughter could and would be independent was thus a political statement of no small importance. This was especially true as the Condorcet family had deep aristocratic roots. By this phase in the Revolution, aristocrats—even those who supported the new regime—encountered a great deal of suspicion. For Condorcet to adopt a rhetoric of self-sufficiency, independence, and quiet virtue was a timely choice for an aristocrat, and further underscores the revolutionary character of this document. Comparing Condorcet's priorities with those of Du Châtelet, who raised her daughter several decades before, likewise reveals how far Condorcet had moved from the traditional goals of the nobility. Du Châtelet's main concern was to raise her daughter to be a bride, one whose prestigious marriage would bring honor to her family; Condorcet had more individualistic and independent aspirations for his daughter's future.

Apart from his emphatic belief that Eliza should pursue occupations that contributed to her independence, Condorcet remained quite open on the subject of what she should study. Her natural talents and facilities had not yet made themselves clear, and so it was difficult to say what activities would be most pleasing and productive.[113] Regardless of the occupation to which his daughter chose to apply herself, she should find consolation in her work and not be dependent on other people for her happiness: "if you cannot live alone, if you have need of others to escape from boredom, you will find it necessary to mold yourself to fit their tastes, their wishes, and they might by chance take away from you these means of filling the voids of your time, when they no longer depend on you for anything."[114] The dreaded dependence came in many forms.

Without reference to God or religion, he urged his daughter to cultivate her sensitivity and charity, to aspire to relieve the needs of the poor and the miserable. True happiness resided in the connection to and caring for humanity that he had counseled: "You will find . . . that it is sweeter and more useful, if I might dare to say so, to live for another. It is the only way that you yourself can truly live."[115] Independence was one thing; selfishness was something else entirely.

Unfortunately, Condorcet was unable to direct his daughter's education. Worried that his presence endangered the innkeeper who was hiding him, he attempted to escape Paris in 1794. He was soon apprehended and died in prison. His letter to his daughter, written in the months before his escape, was his last effort to guide her education. Condorcet was not the first father to encourage his daughter to perform works of charity or to cultivate a secular form of virtue. His exhortations to independence, however, were extraordinary. He imagined a life for Eliza outside the constraints of feminine dependence. Regardless of her financial or marital status, he asked that she remain independent and in control of her own life, a revolutionary idea for an eighteenth-century father.

Condorcet is sometimes criticized as an abstract and impractical thinker, but his plans for Eliza show that he was capable of balancing grand schemes such as his national education system with small-scale individual reforms. Like the other savants discussed in this chapter, he utilized his child's education as a way to implement some of his most cher-

ished ideas and to reform society from the bottom up. His plans for a new
revolutionary system of education may have stalled, but he could still see
his ideas put to use by ensuring that his child was raised according to his
principles.

These various educations and educational schemes show that Enlightenment thinkers found girls' education an interesting but vexing subject.
They evinced contradictory ideas on the subject. Du Châtelet wanted her
children to be sufficiently "enlightened" and polished so they could easily climb noble hierarchies. Diderot and D'Epinay believed their daughters
to be lively, intelligent, and virtuous—quite unlike the passive Sophie of
Rousseau's *Emile* or the manipulative, frivolous little girls that pedagogical theorists criticized. Enlightened parents could be innovative in the
ways they educated their children: nature walks, conversation, teaching
them relatively new subjects like anatomy while still grounding them
in traditional disciplines such as history. The key goal was to develop a
child's ability to usefully contribute to society and to live as a virtuous
citizen. For girls, utility generally entailed reproductive and domestic labor, and their educations reflected that. When pondering their futures,
however, contradictions set in. Although some thinkers recognized the
constraints that marriage placed on women, they still found the civic virtue of wives and mothers appealing and essential. The pull of the latter
proved irresistible to most men and women of letters. Only Condorcet exhorted his daughter to remain independent, even if she did marry. Others
were more willing to accept a trade-off: civic virtue and respectability in
exchange for autonomy.

## CONCLUSION

For philosophes, the family functioned as a sort of laboratory, a site where
thinkers could apply new pedagogical methods and ideals of utility and
civic virtue. They avoided methods of memorization and recitation, they
adopted a largely secular understanding of morality, and they enshrined nature as an ideal classroom and blueprint for society. Many aspired to raise
their children as self-effacing citizens who would be renowned for their virtue. Rather than working to attract attention for their unusual erudition,
these parents hoped their children would blossom into competent, practical, and talented citizens of the *patrie*. The family became a locus for on-
the-ground reform, allowing parents to change society one child at a time.

The thinkers discussed here published their ideas about education and,
as they had in inoculation debates, incorporated stories from their own ex-

periences. In fact, eighteenth-century pedagogical literature might be the most emphatic genre of the age in terms of authors stressing verisimilitude. Domestic life and especially the education of children offered men and women a key venue for self-fashioning. In addition to outlining their theories of human nature, progress, and social reform, they also modeled themselves as citizens raising a virtuous generation. They did not have to make this argument explicitly, a strategy that could have seemed overreaching and arrogant to some readers.[116] Yet by stressing how much care they took with their children's education and how delighted they were by their offspring, philosophes could show their public that they loved their children very much and worked hard to secure a happy and virtuous future for them. Writing about their children's education enabled men and women to imply that they were paragons of sentimental affection and public-minded citizens. They displayed one method for cultivating a moral society and turning children into citizens.

This genre proved particularly popular with women of letters. The subject matter corresponded to the most virtuous and respectable activity for women: raising their children. Women who devoted themselves wholly to their children's education were not strange or distressing figures; instead, they seemed to be that most virtuous incarnation of femininity, the woman fully attuned to nature's dictates and her children's needs. The figure of the devoted mother was ubiquitous during the age of sensibility: clinging to her children in portraits, modeling the joys of motherhood in sentimental novels, nursing her children while her ecstatic husband looked on. In fictional representations, these good mothers demonstrated their virtue by their willingness to sacrifice themselves: to deny themselves pleasures or even die for their children's benefit.[117] But in real life, there was a bit more room to play. Women like Émilie Du Châtelet and Louise d'Epinay could use their maternity to spotlight their personal virtue and to shepherd their ideas into the public.

Philosophes' education of their children thus worked on multiple levels. On the one hand, these documents suggest that men and women of letters used their children's education to practice fashionable pedagogical ideas. And by publishing seemingly true-to-life accounts of these educations, philosophes suggested their own virtue and sensibility by reminding their readers how assiduously they worked for their children's benefit. They reaped both private and public benefits from the practice and representation of education.

# CHAPTER FIVE

# Organic Enlightenment

Antoine Lavoisier, best known for his work on the chemical composition of air, left behind several portraits of himself. These portraits tend to show him indoors: directing experiments in his laboratory, gazing at the viewer, or collaborating with his wife Marie-Anne. These are the images of Lavoisier that we know best. In 1788, however, he argued that savants must also pursue their work outside the confines of the laboratory. "It is not only in our studies that we must research political economy," he insisted, "[but also] through a reflective study of a great exploitation of land, by calculating the distribution of rising wealth over a large number of years. That is how we can develop sound ideas as to what contributes to the prosperity of a large kingdom."[1]

When Lavoisier exhorted his fellow philosophes to get out of their armchairs, leave their studies, and start working in the world, the subject at hand was agriculture and his audience was the Société d'Agriculture. Although he is better known for his work on oxygen, Lavoisier also studied agronomy. He exhorted a generation of savants and wealthy landowners to experiment with new agricultural techniques, to renew French agricultural production, and to make farming a noble and patriotic pursuit. His efforts on this front reveal how one philosophe viewed his role in the world and how he worked to transform his corner of the kingdom into a showroom for Enlightenment reform.

By the time he read his paper to the Société d'Agriculture, Lavoisier had already invested a fortune in agricultural experiments. He had patched together various tracts of land in the Blois region: 127 hectares in Villefrancoeur, 694 hectares in Fréchines, and so on. His holdings eventually totaled 1,129 hectares. The bulk of his estate was at Fréchines, where Lavoisier built his country château.[2] Fréchines and its adjacent lands constituted an experimental farm, a site where Lavoisier tested agricultural techniques such as growing new crops and introducing new breeds of livestock.[3] His goal was to increase the productivity and profitability of his farm. Making Fréchines more fertile would provide an example for other farmers to follow as they sought to revitalize their own holdings.[4]

This experimental farm might seem to bear little relation to the domestic experiences discussed in the previous chapters, but both were undergirded by shared assumptions and ideals. In eighteenth-century France, the family was a popular social metaphor. Society was inherently familial and, more specifically, patriarchal. Even though Lavoisier did not have children, he still enacted a paternalistic social vision on his farm. He conceived of the farm as more than a laboratory in which he would generate data: he also relied on Fréchines to cast himself as an ideal enlightened patriarch and zealous public reformer. He sought to model the ideal benevolent landlord who looked after his neighbors and tenants, loaned money to help them through lean times, and acted as a moral exemplar for them to follow. He had a holistic social vision for Fréchines, one that went beyond planting turnips. He founded a new school, made charitable gifts, and intervened in family squabbles. Lavoisier's experiences at Fréchines make for a rich example of Enlightenment paternalism and social reform outside the context of biological paternity. Although Lavoisier's neighbors were hardly children, much less his children, he credited himself with considerable power to improve their lives.

As the years wore on, Lavoisier increasingly used the farm as a backdrop for his public image. When he needed more data to suit the ends of agricultural reform, his lands at and around Fréchines helped him gather that information. When he performed the role of the ideal landlord, he did so at Fréchines. When he wished to demonstrate his patriotism and commitment to the Revolution, he called attention to his long-term residency at Fréchines and the good works he had performed there. He adopted many of the rhetorical techniques used by savants in domestic contexts: writing about his personal experiences, stressing his emotional bonds, and emphasizing his personal virtue and suitability for public life. Accordingly, Lavoisier's efforts provide an opportunity to examine the sentimental savant in action, this time outside the domestic sphere.

Unlike my previous discussions of Lavoisier, which have highlighted the close collaboration between Marie-Anne and Antoine Lavoisier, this chapter focuses more exclusively on Antoine's work at Fréchines. Marie-Anne spent time at Fréchines, but she preferred the diversions of Paris and tended to stay there. For that reason, her role in her husband's agronomy

work was smaller than it was in other areas of his life. Their marriage
nevertheless remained crucial to their public image. From a public relations perspective, their relationship looked the same, even if they were
separated more often. Marie-Anne continued to craft Antoine's public image, this time as an enlightened landlord, and Antoine continued to fashion himself as patriotic, learned, and sensitive, both with and without his
wife's help.

The story told in this chapter stretches from Lavoisier's purchase of
lands at and near Fréchines through the years following his death, when
his memorialists called attention to his work on the estate as evidence
of his intellectual rigor and paternal benevolence. Most of the documents deal with Lavoisier himself: the story he told about Fréchines in
his correspondence and publications, and the ways in which his eulogizers continued that story for him after he died. Although I would like to
know more about the other side of this story—what Lavoisier's neighbors
thought about his work, how patriotic they deemed him to be—such an
examination is largely outside the purview of this chapter. My focus is on
Fréchines as Lavoisier imagined it, as a place where he invested his time
and money in order to pursue his intellectual goals and manage his public
reputation. This chapter is thus an examination of how Lavoisier crafted
an image of himself as the ultimate enlightened reformer.

## RENEWING THE LAND, REVITALIZING THE NATION

The fundamental purpose of Fréchines was to gin up support for new agricultural methods. Agriculture increasingly featured in proposed economic
reforms and became a much-admired industry in eighteenth-century
France.[5] The moral significance of agriculture is evident in François de
Fenelon's *Adventures of Telemachus*, a popular 1699 story of Ulysses's
son and his wanderings. Guided by his companion Mentor (the goddess
Athena in disguise), Telemachus encountered many different rulers and
styles of ruling. In this popular and influential treatise on politics, agriculture served as a bellwether. If a land had a flourishing agricultural sector—
defined as leaving no lands uncultivated and maximizing the produce of
those lands under cultivation—the travelers praised the ruler of that land.
Wasted lands and idle people, however, reflected poorly on their sovereign.
A successful leader encouraged the industriousness of his people, particularly in the all-important sphere of agriculture. More than crops flourished
as a result of wise management: marriages and morals also thrived. As
Mentor explained to one king, "Thus will your whole territory, in time, be

peopled with healthy, vigorous families employed in agriculture." Even in lands currently underutilized, a successful ruler could encourage progress by ensuring that "the plow will again be held in honor."[6]

In the eighteenth century, many feared that France was falling behind other European nations in its agricultural productivity. French observers worried that domestic agriculture had stalled and that exhausted soil and outmoded methods would only make things worse. Despite these concerns, scholars have demonstrated that peasants and other farmers actually did innovate their growing practices, albeit gradually.[7] Peasants embraced capitalist practices such as specialization, commercial agriculture, and weaving. They were more open to change than eighteenth-century literature allowed. Yet such quiet reforms had little impact on the Enlightenment reading public, which continued to fear that agriculture was failing.

French writers felt especially glum when they compared domestic agricultural production with that of England. Not only did English farmers seem to reap richer harvests; their methods appeared much more modern than those used in France. English farmers had embraced new farming techniques designed to maintain the fertility of their lands, beginning in the early seventeenth century.[8] Cultivators grew root vegetables and leguminous crops as a way to renew tired soil. Planting these crops was efficient, especially in comparison to leaving land fallow. Turnips and potatoes drew the most attention and were used as animal feed (at this time, both crops were considered unfit for human consumption). English agronomists did not stop there. They had many other tools in their kit and aspired to make every aspect of farming more efficient and more scientific. They calculated how much manure, and of what variety, best fertilized crops, which feeds maintained livestock, and which tools made the processes of seeding and harvesting more uniform and less wasteful.[9] The guiding principles of the movement were efficiency and productivity, anchored in scientific methods of study.

In the middle of the eighteenth century, French writers urged these techniques on Gallic farmers. Perhaps the most significant text of this wave of French agronomy was Henri-Louis Duhamel du Monceau's *School of Agriculture*, a book that anchored all subsequent French agronomy; it was the "indefatigable Duhamel" who inspired Lavoisier to pursue agronomy.[10] Duhamel and other French agronomists proposed an agricultural program very similar to the English one: planting root vegetables to renew the soil, collecting more manure to fertilize fields, and raising more livestock to produce that manure. They also hoped to implement English policies of enclosure, arguing that converting commonly held lands to private domains encouraged investment and innovation. Bestowing agricultural honors on effective farmers would have a similar effect, they believed, and so agronomists lobbied government officials to create prizes for agriculture to help improve farming's prestige.[11]

Agricultural societies soon cropped up in various provinces. These societies shared the latest agronomist texts, encouraged their neighbors to adopt new techniques, and bestowed prizes on successful cultivators. To make apparent the benefits of their methods, societies in Limoges, Lyons, and Riom founded experimental farms. These testing grounds helped societies demonstrate the agricultural reforms that they advocated. Because members of agricultural societies depended on government support for their work, they shied away from ambitious schemes to revamp France's financial structures. Instead, they focused on technical improvements and making farming a more honorable profession. Although they did not transform popular agriculture, these societies captured the attention of France's elites, many of whom dreamed of renewing the nation through agricultural improvement.

The intensity of this interest in agricultural reform might be hard to fathom. As John Shovlin has shown, however, the interest in agronomy was not solely about agriculture.[12] Agriculture provided writers with a vocabulary to discuss moral regeneration, enlightenment, population growth, and a return to classical virtue. Talk of renewing soil, revitalizing agriculture, and increasing crop production shaded into discussions of renewing virtue, revitalizing patriotism, and increasing population. Agronomy texts contributed to debates about luxury, social utility, and the purpose of the nobility, all of which were of significant interest in the late eighteenth century.[13] Agriculture assuaged some of the most profound cultural fears of the eighteenth century, a time when many worried that French morals were declining and the French population was plummeting. Moralists feared that flourishing luxury trades were only exacerbating this situation. Seduced by the frivolities of fashion, men were becoming too much like women and women too much like foolish children. How could a nation distracted by such silly pleasures ever bounce back from the embarrassment of the Seven Years' War and develop true civic virtue? This question was of special importance for the nobility, which felt recent military defeats keenly.[14] A flurry of texts urged the French, especially the nobility, to recommit themselves to lives of virtue and patriotism, defined as a greater commitment to the public good than personal interest.[15] Agriculture seemed to provide a way out of all these problems. Agronomists and like-minded authors believed that a renewed commitment to agriculture would lead to a more virtuous, healthy, and prolific France. Population would boom, citizens would devote themselves to cultivating healthy crops and patriotic families, and the national economy—not to mention national morals—would strengthen.

The eighteenth century's love affair with nature and classical virtue also drove the agronomy boom. Living off the land, surrounded by nature's bounty, acquired a new moral significance. Nature provided a model for good behavior, a blueprint that human society should follow. To be "natural" was to be good; to be removed from nature was artificial and debilitating. In an age when many believed that environmental factors influenced character, those who lived closer to nature had more virtue. Rousseau did more than any other writer to popularize the glories of nature, and his paeans to rural life led many of his readers to fantasize about the simple pleasures of farming, even if they continued to live in cities. Although he pinpointed the development of agriculture—and hence the origins of civilization—as the starting point of man's decline from the state of nature, he still saw farming as the best way to live in a corrupted age.[16] Many agreed with him: agriculture seemed a much-needed antidote to the moral rot that infected France's urban centers and especially its royal court. A corrosive taste for luxury and frivolity was eating away at the nation, and writers saw evidence of the damage everywhere they looked. People were spending more than ever before on fashion and simple luxuries instead of devoting themselves to virtuous tasks such as tilling land and raising children.[17] But such contrasts were not the exclusive purview of stern moralists: the very social elites being chided for their rampant consumption and selfishness were charmed by stories of rural and village life.[18] Agrarian festivals seized the imaginations of many Enlightenment figures, who marveled at the simplicity and charm of country life.[19] These stories do not reveal engagement with actual peasants, for a wide social divide separated urban elites from peasants, but instead were fantasies of rural life spun to suit the fancies of city dwellers and courtiers. Nevertheless, agronomists benefited from this enthusiasm, and improvement tracts proliferated beginning in the middle of the century.

Yet readers' fervor for agronomy was little practiced. Duhamel du Monceau, one of the early French agronomists, wrote in his 1759 *École d'Agriculture* that "nothing is more humiliating for humanity than the contradiction that one observes between the opinions and actions of men. . . . It appears that a universal lethargy strikes and numbs our arms."[20] Theory and practice had yet to come together.

Lavoisier's passion for agronomy blossomed in this environment.[21]

As part of the revival of agronomy in the 1780s, he pursued agronomist reforms with great enthusiasm and pondered why progress had been so slow.[22] He worried that agricultural improvement tracts lacked a strong empirical foundation and that, as a result, agronomists did not have clear data supporting their theories. He believed that this shortcoming explained at least some of the reluctance to implement new methods. Agronomists like himself could "perform an important service for the cultivators of this canton by giving them an example of agriculture directed by best principles."[23] An experimental farm like Fréchines, geared toward quantifying the precise impact that new techniques could have on old soil, would solve that problem. Rather than imagining the estate as a rural idyll where he could escape society, Lavoisier called attention to the social utility of his actions at Fréchines.

His example would, ideally, persuade fellow landowners to follow his lead. If Lavoisier wished to bring about a massive transformation in French agricultural practices, he needed to convince proprietors, and especially wealthy proprietors, to bring their practices in line with agronomist principles. He hoped to overcome what he saw as the most significant barrier to reform: a general distaste for farming, especially among the nation's wealthier subjects. Agronomists believed that a major reason elites did not invest in agriculture was because it did not seem like an honorable occupation. Although many members of the reading public were intrigued by agriculture, they fretted that court nobles and landowners did not share their enthusiasm. Lavoisier and other like-minded reformers were anxious to encourage proprietors to have a real impact on agriculture, rather than simply reading about it. By proving that agricultural improvements could substantially increase productivity, Lavoisier hoped to persuade others to follow suit.

Many hurdles stood in the way of this goal, as agricultural experiments were expensive. Even the finance minister Anne-Robert-Jacques Turgot—a figure interested in agricultural reform—had felt obliged to close the experimental farm run by the agricultural society of Limoges because of the high costs associated with running the estate.[24] If Lavoisier hoped to succeed, therefore, he needed collaborators with sufficient capital: "agriculture cannot be improved or regenerated in France except by rich landowners who would be willing to sacrifice some of their profits so as to better cultivate their lands." In particular, they had to have the resources to carry out experiments over the long term, perhaps a couple of decades. Peasants and small landowners did not seem crucial to the success of this project. Instead, agricultural reform would be a top-down process, driven by large estate holders and wealthy patrons. Lavoisier hoped that eventually the cost of reform would decline and become more accessible, but in 1788 he lamented that "ordinary farmers are far from being able to make the necessary changes to undertake such an enterprise, and when they are in a position to do so, I have already observed that they will find it more useful to invest their funds in the capital or in centers of commerce."[25] By advertising himself as an enlightened savant and prudent landowner willing to invest in agricultural reform, Lavoisier hoped to inspire other wealthy subjects to follow his example.

Fréchines thus gave Antoine Lavoisier the means to model a rich, enlightened proprietor who was willing to invest in the nation's future. By demonstrating his willingness to act selflessly and in the public's best interests, he prodded nobles and notables to display their own civic virtue.[26] The virtuous and well-heeled landowner was at the center of his vision of agricultural success.

Lavoisier was not, however, alone in founding a model farm; he was not even the first to do so. Other eminent landowners had founded similar estates, as had the savant Samuel Pierre Du Pont de Nemours.[27] Despite the fact that he was not first in the field, however, Lavoisier argued that each farm made a unique contribution because theories had to be modified to suit local climates. He had to adapt his own theories. For example, he had expected alfalfa to be a good feed crop but had found that it "did not succeed in my lands." Clover likewise proved disappointing, as it only thrived in rainy years.[28] His findings, like his farm, broke new ground.

Although his concept might not have been new, his methodology was. He modified accounting registers to keep track of his expenses and profits for each field in order to quantify each method of agricultural improvement; this practice mirrored his methods in chemistry and in tax farming.[29] Like a good chemist, Lavoisier meticulously measured the seed used and the yield of each crop. "By this method," Madame Lavoisier noted, "we always knew the cost and output of each piece of land and we saw each season the improvements that the right cultivation, manure, and pastures had procured, taking into account the seasons and observations of wind, rain, heat, and cold."[30] Antoine Lavoisier may not have been the first to develop an experimental farm but he did so with exceptional rigor.

Lavoisier soon pinpointed the major problem slowing crop growth at Fréchines: an insufficient amount of fertilizer. A need for more fertilizer meant that the farm required more manure, and therefore more robust herds of livestock to produce said manure. This spotlighted the major difference he perceived between English and French agriculture: the

English focused on raising livestock, and the French on growing wheat. He believed the English approach to be more reliable and efficient. In the English system, livestock ate turnips and potatoes, crops which did not overtax tired soil; in the French system, exhausted land lay fallow. Making the switch to the English system was, however, more complicated than simply adding heads of cattle and sheep to his flock. A proprietor could not simply switch from one model to the other in one season; the transition had to take place slowly. Despite the prevalent assumption that "nothing seems easier to reanimate than languishing agriculture, and . . . that we need only animals and money to do so," Lavoisier's experience indicated that the new animals had to be introduced carefully and only after the farmer had managed to grow enough fodder to maintain them.[31] To obtain more manure, Lavoisier needed to find a way to produce more feed.

His actions on this front were taken right from the handbook of English agronomists: he planted turnips and potatoes. Convinced these crops would do the trick, he spread the root vegetable gospel to his neighbors.[32] To encourage them to grow turnips on their own land, he distributed "seed freely to those who asked me for it." By 1785, he had only "a small provision remaining but this will soon be exhausted."[33] Apparently, turnips proved popular, at least when the seed was free. To further hasten the transition to English agriculture, he also converted some lands to pasture (prairies artificielles) for grazing. Slowly but surely, he transformed Fréchines.

In assessing these experiments several years later, Marie-Anne Lavoisier presented the reformation of the farm as total: "He introduced potatoes, which had been previously unknown. He established pastures where they had never before been. Herds of beautiful livestock filled the farm." In her account, Lavoisier inaugurated methods for agricultural production and transformed his lands. Madame Lavoisier proudly noted that "his success was such that, in the harvest of his fifth year of cultivating the land, the production of wheat had doubled and the farm could now support five times as many livestock as it had been able to nourish previously."[34] After years of careful management, Fréchines operated in accord with agronomical principles.

This account was written years after the experiment had begun, however. Despite his wife's retrospective confidence, Lavoisier himself had had doubts about the potential for agronomy in and of itself to transform the land. Although he remained passionate about new agricultural methods and continued to urge his neighbors to follow his lead, he insisted that his efforts would be ineffective if not accompanied by wholesale political and economic reforms.[35] Farming methods were not the only problem. France's convoluted tax system also discouraged innovation and growth.

By insisting on a two-part solution, Lavoisier sampled ideas from physiocracy, a popular school of economic thought that proliferated from the late 1750s to the 1780s. Physiocrats such as François Quesnay argued that the economy was governed by a set of natural laws and that nature benevolently guided economic growth. It was the job of the *economiste* to discover the laws governing the economy and obey nature's dictates.[36] Physiocrats came to the conclusion that agricultural development was the best way to grow the economy. Unlike all other trades, they argued, agriculture was the foundation of economic growth; the government's current focus on manufacturing and international commerce as drivers of the economy was wrong. In advocating agriculture as the only source of renewable wealth, the physiocrats broke with the mercantilist school of thought. Physiocrats insisted that the economy would fare better if the government would free the grain trade, rather than maintaining an artificially low price for bread. They trusted that these reforms would happen if wealthy proprietors accrued greater control over France's land. Peasants, they believed, were ineffectual farmers who resisted and slowed reform. If France would move away from peasant farming to commercial farming, reforms would follow.[37] Although they both saw agriculture as central to the future of France, agronomists and physiocrats did not always agree. Physiocrats focused on the political processes that impeded the growth of natural wealth; agronomists paid closer attention to techniques of farming and planting and ignored the larger economic circumstances.[38]

When Lavoisier arrived on the agricultural scene, he showed little interest in taking sides. Instead, he borrowed from both agronomy and physiocracy. He followed the techniques espoused by agronomists: planting turnips and potatoes, cultivating livestock. Yet he also found the *laissez-faire* approach of the physiocrats compelling, as made clear by his frequent praise for his friend the physiocrat Samuel Pierre Du Pont de Nemours. Physiocrats lobbied the government to lower the taxes associated with agricultural production, reform France's convoluted system of overlapping privileges, and simplify taxes overall. Lavoisier agreed. In his correspondence and publications, he noted the impediments posed by the French system of taxation, namely, the assessment of the *taille*. This, the most substantial of all French taxes, was calculated according to the heads of livestock that a given farm could support. This was a problem, given the emphasis that Lavoisier and his fellow agronomists placed on livestock and their manure as key to renewing and sustaining agricultural productivity. He considered livestock to be so important that he offered financial incentives to his tenants if they would increase their herds. This incentive had little effect, however, because any financial gain offered by Lavoisier would be offset by the increased tax burden. The solution, therefore, was that the tax system needed to be simplified.[39] Lavoisier continued to pursue a program of agricultural improvement even as he pushed for political and economic reforms. His methods at Fréchines were thus an ecumenical mix of agronomy and physiocracy. His actions and writings on agriculture shifted fluidly from one genre to the other.

Lavoisier was not alone in his desire to reform the tax system. Taxation was a heated political issue: nearly everyone believed that they were overtaxed and that other groups were undertaxed. Bedeviling matters further was the inchoate nature of French taxation.[40] The government levied a vast array of direct and indirect taxes; these were complicated by the series of overlapping privileges issued to guilds, cities, and social orders. These privileges had been designed to ensure the loyalty of the most powerful members of society, and they became central markers of nobility.[41] Bestowing privilege constituted a well-established political strategy in early modern France, and monarchs continued to confer privileges of all sorts until the practice was abolished during the Revolution. Although many contemporaries recognized problems with this system, reform proved difficult to realize. The most privileged members of society—the nobility and clergy—believed themselves to be shamefully overtaxed and they resisted any proposal that involved their taking on new fiscal duties. Members of the nobility and clergy did not have to pay the *taille*, the direct tax on land, but they did contribute via the *capitation* and *vingtième* (both direct and universal taxes). Many members of the First and Second Estates believed these taxes undermined their social prestige; as a result, they were reluctant to agree to pay more.

Yet as minister after minister argued, the French state was in dire need of revenue. The state's debt had reached perilous levels, especially after the borrowing sprees that had funded the Seven Years' War and the American Revolution. By the eve of the French Revolution, fifty percent of all state revenue was being siphoned off to pay interest on outstanding loans. Desperate to improve its finances, royal ministers proposed a flurry of reforms, even convening representative bodies such as the Assembly of Notables (which had not met since 1627). Yet these efforts at reform were all thwarted, undermined by the lobbying efforts of privileged groups and widespread mistrust of the absolutist state.

Although he was aware of these political difficulties, Lavoisier also lobbied for reforms to the *taille*. As the years progressed, Lavoisier held out hope that the government would make agricultural investment a permanent part of the budget, freeing agronomy from the vagaries of political negotiations.[42] He also urged the government to alleviate tax burdens, which he believed stifled innovation and investment. Only by reforming the tax system could farming become a more profitable and attractive profession; until then, agriculture would continue to stagnate.[43]

The empirical value of Fréchines therefore extended beyond agronomy. Lavoisier used the farm to argue in favor of tax reform as well as new methods of cultivation. He geared his model farm toward different audiences: fellow agronomists, wealthy landowners, and government ministers. Lavoisier's vocations—as a tax farmer, chemist, and gunpowder manufacturer—made him particularly adept at navigating these different professional settings. Having honed his empirical and administrative talents elsewhere, he was poised to use Fréchines effectively. It became a key component in his campaign of persuasion, a place to generate proof and inspire emulation.

In this, as in other matters, Lavoisier proved himself to be a skilled manipulator of eighteenth-century intellectual ideals. He avoided the appearance of ideological purity and instead stressed his pragmatism and adaptability. Many savants practiced this sort of "sensibilist science," making them loathe to pursue abstract systems. Labeling a thinker "a builder of systems" was a choice insult.[44] By picking and choosing between agronomist and physiocratic solutions, Lavoisier allied himself with the sentimental approach: to be wary of systems and ready to adapt to particular circumstances. He emphasized a blended approach and stressed that his choices were commonsensical.

Sensibilist savants also insisted that virtuous emotions, rather than sterile rationalism, should guide intellectual inquiry. In keeping with this ideal, Lavoisier emphasized the paternal sentiments that motivated his work at Fréchines. He represented himself as caring deeply for his tenants, neighbors, and, indeed, all of France. Anxious to improve their lot in life, he helped them learn about new farming techniques and lobbied on their behalf for political reforms. He saw himself as a benevolent and selfless father figure.

Paternalist ideals had, of course, long circulated in France. The nobility used vocabularies of merit, paternalism, and industriousness to define and defend their role in society.[45] In the second half of the eighteenth century, provincial nobles amped up such rhetoric. As it became fashionable to lampoon luxury as wasteful, morally corrosive, and emasculating, the schism between court nobles (whose lives as courtiers provided much grist for the antiluxury mill) and country nobles widened. Provincial nobles denounced luxury and insisted that they—unlike their profligate court counterparts—invested their money wisely and effectively. They wrapped themselves in the cloak of agricultural improvement. In so doing, they appealed to ideals of merit and republican morality to justify their lofty social standing and to present themselves as improving the lives of those around them. Aristocrats had long used such language of merit to explain their social value, but it was a rhetoric easily co-opted by nonnobles. By the end of the eighteenth century, it was difficult for nobles to claim they enjoyed a monopoly on patriotism, merit, or civic virtue. It became possible for other social groups to adapt erstwhile noble rhetoric for their own purposes.

Extensive debate of what constituted a "good father" further underscored the moral value of paternalism. As I discussed in chapter 3, sentimental paternalism was much in vogue, and the archetype of the *bon père de famille* appeared in countless political tracts, plays, novels, and petitions. A good father was one who ruled through kindness and affection, not fear and coercion. He set a positive moral example for his wife and children and encouraged them to emulate his good behavior. The appeal of the concept was not limited to biological fathers but instead played into paternal metaphors of all sorts. Eager to distinguish himself from his unpopular grandfather, Louis XVI styled himself as this new sort of father, both to his children and to his subjects. Anyone claiming paternalistic dominion over others could be likewise influenced to present himself as a modern, sentimentalized figure, not a stern authoritarian.

Lavoisier borrowed from these debates. Like a reform-minded nobleman, he was qualified to lead by his personal merit, good morals, and paternal tenderness. Like a good father, he would carefully and sensitively guide his neighbors to make the "right" decisions. And like a good savant, he emphasized his sensibilist bona fides. Lavoisier's superior learning and good morals made him an ideal candidate to educate and enlighten his neighbors (or so he thought). He would train them to improve their farms and care for them in times of need. Lavoisier's vision of himself thus matched the self-representations of provincial nobles, although he added a sentimental and scientific twist.

One way that Antoine Lavoisier demonstrated his paternal beneficence was through dispensing charity and financial assistance. Sometimes his assistance proved insufficient to arrest the decline of a particular farm or town. Even then, however, he stressed his emotional connection to his less fortunate neighbors. As Marie-Anne Lavoisier wrote, he "fulfilled his administrative duties with the zeal for the sweetest philanthropy and if the laws forced him to act with perhaps too much severity, he always attempted to soften the effects."[46] Eager to help, reluctant to punish, Antoine Lavoisier presented himself as a paternal landlord in every way.

This emphasis on helping the needy corresponded with late eighteenth-century ideals of charity. Charity had once fallen under the purview of the church, with individual donors most often motivated by their religious beliefs. In the second half of the eighteenth century, however, a newly secular approach to charity developed alongside the older Christian model. Lavoisier's interventions fit within this worldly framework. When discussing his charitable endeavors, Lavoisier did not mention God or religion. Instead, he focused on the feelings and social bonds that motivated him.

In caring for his fellow citizens, Lavoisier modeled himself as a good patriot as well as a good savant. The ideal savant was selfless, devoted to the common good, and willing to risk life and limb. The same could be said of the ideal patriot. Lavoisier's representation of himself as savant and citizen overlapped in key ways. His desire to present himself as an active patriot drove his work at Fréchines. To prove his patriotism and his intellectual engagement, he spent time in the world solving everyday problems. It was, of course, still admirable for savants to devote at least some of their time to quiet study—and indeed, Lavoisier himself was keen on such practices.[47] Too much isolation, however, helped no one, as I discussed in chapter 1. At some point, it became necessary for savants to put their books aside and step out into the world. An approach balanced between reflection and action seemed to provide the best path forward to social reform.

One of Lavoisier's greatest challenges, the dry spell of 1785, provides a case study for his attempt to pursue a path of pragmatic but benevolent reform. Low rainfall threatened French crops, farmers' livelihoods, and Lavoisier's progress. The disaster produced a "state of distress" in the region: there was a "near total failure for the alfalfa harvest," which left Lavoisier with "no choice for this year but to plant buckwheat." At least he had the capital to persevere. He noted with sadness that "two of my farmers have been obliged to yield to the unhappy circumstances; they have declared bankruptcy. I had no choice but to seize their harvest and to have their land worked at my own cost and seeded with wheat for this year; if not, they would have lain fallow."[48]

Although he did seize land from farmers unable to pay taxes, the dry spell also spurred Lavoisier to charitable action. Concerned by the poor conditions for wheat, he wrote, "[I] took it upon myself to send straight-

away for four hogsheads or forty-eight *septiers* (measure of Blois) of buckwheat from Sologne, where the cultivation of that crop is common. I will
receive them sometime after tomorrow and they will be distributed to the
farmers of these two parishes in proportion to the number of acres they
farm."[49] He hoped that the hardy crop would succeed and help stave off
bankruptcy and starvation in the region.

These actions speak to Lavoisier's pragmatic adherence to physiocratic
principles of free trade. Although he wanted to free the grain trade, he did
not want people to starve. When the government intervened in the crisis,
he lauded its charitable intentions. He expected government officials to
balance free trade ideals with the human cost that weather and natural
disasters could inflict. He adhered to liberal economic principles, which
he mentioned to municipal officials in Blois in 1789: "The question of
whether or not the government should interfere with the commerce of
wheat is an important one; those who have reflected most on this issue are
convinced that an absolute liberty is sufficient to establish an equilibrium
of prices in different provinces." But he also cautioned that "the alarm is
being sounded in the Assembly of the three orders reunited; it will soon be
transmitted to the most indigent class of people. . . . To put it succinctly,
the people are becoming alarmed, the poor are suffering; this is not the
moment to argue, it is a time to give aid."[50] Lavoisier proved more than
capable of balancing his intellectual ideals with pragmatism and compassion. Otherwise, he worried, popular unrest and human suffering would
have devastating consequences before the market could stabilize itself.

Lavoisier's actions during the dry spell of 1785, as well as the other
items on his Fréchines agenda, demonstrate his desire to represent himself
as an ideal savant and ideal patriot at the same time that he reformed the
agricultural production of the region. He presented himself as pragmatic,
charitable, and deeply concerned about his fellow residents—an empiricist
but also a social reformer. In short, he fashioned himself as the perfect Enlightenment savant and the perfect landowner. Here was a clear example
of a philosophe applying new knowledge to solve old problems.

As the backdrop to these self-representations, Fréchines provided Lavoisier with an authority both empirical and moral. He used the estate
as a way to display himself, and not just agriculture, in a flattering light.
To emphasize just how virtuous his actions were, he stressed his personal
and financial sacrifices. For example, he wrote: "The work with which I
occupy myself has already cost me nine years of care and toil; it has demanded great expenditures on my part, expenditures for which I could
never hope to be compensated. . . . But it has taught me great truths that

even the most learned persons only perceive in a vague manner." Although experimental farms cost a great deal to maintain, he considered his money well spent because his financial sacrifices yielded empirical returns. His commitment to the truth was of greater concern than his monetary gains. He also mentioned in his correspondence the generous, interest-free loans he made to nearby towns, again reinforcing his self-representation as committed to the well-being of area residents. He underscored that he did all of this for the benefit of the public: "It led me to hope that one day I will be able to contribute to national prosperity in shaping public opinion through writings and by example."[51] Lavoisier's unsubtle reminders about the financial risks he had taken in cultivating Fréchines drew attention to his virtue and selflessness.

Lavoisier enjoyed a substantial fortune, and so his willingness to sacrifice profits might not seem especially impressive. The time he spent at Fréchines may have exacted a different toll on his personal life, however. Marie-Anne Lavoisier cared little for her husband's country château and tended to remain in Paris during her husband's stints in the countryside. She spent an increasing amount of time with Samuel-Pierre Du Pont de Nemours and, around 1781, became romantically involved with him. Their affair lasted for several years. The Lavoisier marriage, as noted in chapter 2, remains something of a mystery. Marie-Anne and Antoine seemed quite affectionate toward each other, and they shared a deep commitment to chemistry and to promoting their family's reputation. Yet theirs may not have been a heady romance. Given the lack of evidence, it is impossible to know the nature of the Lavoisiers' feelings for each other or how these may have changed over time. Nevertheless, it is possible that Antoine's newfound enthusiasm for agriculture and his willingness to spend considerable time in the countryside encouraged his wife to pursue an affair. Fréchines cost Lavoisier dearly, perhaps in more ways than one.

## "USEFUL WORKS"

Despite her affair, Marie-Anne and Antoine remained an effective team, working together to promote their family's reputation and to fashion Antoine as a dedicated and learned patriot. Madame Lavoisier made clear that Lavoisier's time at Fréchines was not devoted exclusively to buckwheat, turnips, and manure. Eighteenth-century individuals, Lavoisier included, viewed agriculture as just one component of an organic program of social reform, including moral regeneration and population growth. Lavoisier utilized Fréchines to generate agricultural data, but he also saw the estate as something grander. He used his time on the farm to perform the role of the enlightened patriarch *par excellence*: a benevolent, paternal reformer seeking to improve the lives of his fellow citizens, even at great personal cost. He made Fréchines into more than an experimental farm. He turned it into a social laboratory to demonstrate proposed social and economic reforms and to display himself as a model philosophe with the talent and the drive necessary to implement them.

Fréchines thus became an enlightened oasis that featured ideas dear to Lavoisier's heart: public education, charity, and strong family ties. Contemporary sources for these everyday activities can be difficult to find, as Lavoisier focused on his crops and livestock in presentations to agricultural societies. Descriptions of Fréchines written by other authors, however, stressed that Lavoisier's reforms did not stop at the field or the barn. In particular, eulogies written upon Lavoisier's death depicted Fréchines as a showcase of Enlightenment reform and evidence of Lavoisier's good heart. These memorializing texts do more than provide evidence as to what Lavoisier may have done on his farm; they also show that his friends and family considered his work at Fréchines to be a particularly good example of how kind, helpful, and committed to reform he had been. It provided a golden opportunity for fashioning his reputation. By emphasizing his selflessness, social ties, and kindness, memorialists reinforced Lavoisier's representation of himself.

Marie-Anne Lavoisier composed a particularly suggestive narrative of Lavoisier's life. She titled her text the "Notice Biographique de Lavoisier" and wrote it to help Georges Cuvier (then acting as perpetual secretary of the Academy of Sciences) compose an *éloge* for Lavoisier. In describing her husband's many intellectual accomplishments, she noted that his experiments at Fréchines had provided necessary evidence supporting agricultural reform. In particular, she praised his careful accounting methods and his novel techniques for quantifying the gains and losses of his crops, which endowed agronomy with a scientific precision it had previously lacked.

He had done much more than experiment with new crops, however; his time at Fréchines was often occupied by the "many useful works that he could be seen doing amid all the inhabitants." These "useful works" were many and varied, and even included Lavoisier "acting as a magistrate of the peace to reestablish good relations between two neighbors, or to return a son to paternal obedience; in general, giving an example of all patriarchal virtues."[52] Madame Lavoisier's description of her husband calls to mind a highly idealized depiction of rural life: the kind and rational landlord surveying his environs, wandering among his neighbors, and solving any and all problems that threatened to tear the social fabric. Dealing with squabbling neighbors and headstrong sons might seem removed from agricultural experiments, but Lavoisier understood such work as central to his mission at Fréchines. The farm was more than a place where he experimented with new agricultural techniques: it was also where he played the part of the ideal enlightened patriarch whose reason and benevolence equipped him to solve all manner of social problems.

Marie-Anne Lavoisier further claimed that Antoine Lavoisier could even be found "caring for the ill by visiting them, taking care of them, and exhorting them to patience and hope." Nursing the sick was a task that might traditionally fall to the local curé, but the paternalistic Lavoisier apparently took it on himself to care for his neighbors in times of sickness. He also provided charity, "leaving goods just outside the market to preserve the pride of those residents whose shame kept them from seeking aid, which only made them more pitiable."[53] These actions were simultaneously charitable, aristocratic, sentimental, and patriotic, ideals that were intricately connected to each other in the late eighteenth century.[54] Lavoisier acted as the archetypical wealthy man caring for the poor, the local landowner looking after his tenants, the good man driven to action by the needs of the sick.

Lavoisier did not simply cut a noble and moral figure but also worked to make a more permanent mark on the area by enlightening institutions as well as individuals. Like many savants, Lavoisier considered education to be a key component of social reform and believed that a system of secular, public education would help form stronger, more virtuous citizens. Anxious to see this process begin on his own lands, Lavoisier founded a secular school on the estate "for the present generation."[55] He hired a new teacher for the school and paid him a yearly salary of 400 *livres*. The local curé apparently resented the competition for students, which suggests that a significant number of people chose to attend Lavoisier's school rather than learning their lessons from the village priest.[56]

Marie-Anne Lavoisier emphasized her husband's total presence on the estate. He seemed to her the perfect patriarch, one who committed his time and money to charity and public service. In her telling, Antoine Lavoisier devoted himself fully to his fellow residents' well-being and felt a deep emotional attachment to them. He worried little about personal profit or praise and focused instead on others. These were all key attributes of the ideal philosophe and the ideal citizen.

Marie-Anne Lavoisier's sentimental account of her husband's time at

Fréchines was supported by Lavoisier's other memorialist, his collaborator Antoine François, comte de Fourcroy. Like Madame Lavoisier, Fourcroy noted the empirical and agricultural value of the estate. "Agricultural experiments," he wrote, "occupied Lavoisier for nine years. His work titled the *Richesses Territoriales de la France*, for which he amassed the material over a long time, should place him among those writers most worthy to enlighten nations about their true interests." Also like Madame Lavoisier, Fourcroy made clear that Lavoisier's work at Fréchines had a charitable bent: "Apart from the paternal welcome and the services of every type that he gave to less fortunate youths, he also helped them if their tastes and dispositions inclined them to pursue a career in the sciences. Without pomp and without comment, he assisted a host of unfortunates."[57] Fourcroy highlighted similar themes to those included in Marie-Anne Lavoisier's "Notice Biographique": a commitment to improving education for the next generation and to caring for the neediest members of the community.

Fourcroy imagined Lavoisier's presence in the region to have been beneficial and universally appreciated. The younger man's heart ached to think of how area citizens had reacted to Lavoisier's untimely death in 1794. In particular, Fourcroy contemplated "the residents of several communes in the department of Loir et Cher, where [Lavoisier] owned land, [who] will preserve for a long time the memory of his charity and his active humanity. How many times have the refugees of indigence and sadness seen him with his worthy companion! How many tears have they shed together!"[58] Love, poverty, and tears: this passage hit every sentimental high note. Fourcroy imagined that Lavoisier's former neighbors, touched by his efforts to help them, would have remained grateful to him. Lavoisier appeared to be the perfect paternalistic reformer.

As his eulogizers demonstrate, Lavoisier imagined Fréchines as a site for all manner of experimentation. His ambitions for his holdings went far beyond agricultural improvements. His experimental farm was also a social laboratory where he could demonstrate for the rest of the nation how they should run their farms, their families, and their schools; in other words, it was a site where he modeled sentimental paternalism. These representations served two purposes: they highlighted Lavoisier's ideas for a wider public and they polished his reputation.

These texts echo a eulogy composed by Pierre Samuel Du Pont de Nemours for himself, in which he immodestly intoned: "Here lies Pierre Samuel Du Pont . . . [who] retired to this rustic haven, wonderfully assisted by Nicole Charlotte Marie Louise Le Dée, his amiable, active, prudent, benevolent, and beloved wife, where he has built a country retreat, planted trees, introduced the cultivation of new plants, provided work and lessened poverty, and cured the inhabitants of their illnesses."[59] Du Pont's good deeds strikingly resemble Lavoisier's: establishing a country estate, planting new crops, alleviating poverty, and tending to the sick. Du Pont and Lavoisier were not satisfied with advertising agronomy reforms. They also wanted to spotlight themselves as men of virtue who performed good deeds. Their work extended beyond field and barn to include healing, charity, and general benevolence.

The appeal of playing an enlightened patriarch ran deep. A particularly famous example was Voltaire, philosophe extraordinaire and patriarch of Ferney. In purchasing his Swiss estate, Voltaire had been partly motivated by concerns for his personal safety. He did, after all, have quite the knack for landing himself in trouble with any number of grandees and monarchs. Switzerland provided him with a safe haven, but security was only one of the benefits the estate proffered. Ferney and its surrounding lands soon fired up Voltaire's passion for social reform. He found the estate to be in poor shape, with much land left uncultivated and residents devastated by the rapacious actions of tax farmers. The sterility of the people matched the sterility of the land, as Voltaire claimed that his new neighbors suffered from a dearth of marriages and births.

Troubled by this social and agricultural stagnation, he vowed to reverse the situation. Ever confident in his abilities, Voltaire took a hands-on approach to fixing Ferney. He walked among his peasant neighbors, forming personal bonds with them. He provided charity when it was needed. And, in utmost paternal fashion, he oversaw private and public affairs and endeavored to form and strengthen familial bonds. He felt strongly that his presence was required to improve the dire situation at Ferney. As he wrote to a correspondent, "You philosophize at your ease, but I must visit my smallholdings. I must heal my peasants and my cattle when they are sick, I must marry off the girls, and I must improve abandoned fields."[60] The social vision articulated here was paternal and hierarchical. Voltaire's ministrations were essential to the well-being of his peasants and cattle (tellingly grouped together in his phrasing). Apparently, neither social nor agricultural regeneration would occur unless Voltaire personally oversaw marriages and assessed his fields; he saw himself as the driving force of reform.

Voltaire's actions at Ferney added a new social dimension to his moral crusades. The plight of peasants had not been one of his major concerns until he moved to Switzerland. His arrival further developed his interest in social ills, such as the miscarriage of justice in the Calas affair (discussed in chapter 3). He also became fully convinced that the true philosophe would focus on improving the lives of those around him. Accordingly, he devoted much time and money to remedying the ills suffered by his less fortunate neighbors: "I see around me the most frightful misery, in the midst of a smiling countryside. I put on a show of trying to remedy some of the evil that has been done over the centuries. When you find yourself in a position to do some good in a half league of countryside, that is truly honorable."[61] Ever the master of reinvention, Voltaire had found a new starring role: that of the enlightened and benevolent landlord.

His ambitions to help peasants did not stop at the border of his own estate. Mortified by the persistence of feudalism in Mont Jura, a French region close to Ferney, Voltaire campaigned fiercely on behalf of the serfs. Although he did not succeed in abolishing feudalism in the region, the episode became a part of Voltairean lore. Voltaire's memorialists did not present this work as motivated by abstract economic principles but rather by a love of humanity and a desire to protect the downtrodden from life's vicissitudes. The biographer Louis-Pierre Manuel, who lauded Voltaire's efforts and mistakenly believed that Voltaire had ended feudalism in the region, discussed the Mont Jura crusade at the end of his article on Voltaire. He spotlighted Voltaire's paternal benevolence and how deeply the Mont Jura serfs loved their benefactor. "His vassals could only applaud him as their seigneur," he declared. In addition to working to end serfdom, Voltaire had revitalized the economy: "he built them homes, workshops, and a village where he had previously found dozens of poor people, covered in sores and scrofula; now it is well populated with well-off laborers and good artisans."[62] Manuel's article captured Voltaire's ideals and planned reforms, which were a clear predecessor to Lavoisier's plans for Fréchines. In both cases, a paternalistic savant, moved by his inner goodness to alleviate the suffering of the poor, purchased a new estate where he revitalized agriculture, dispensed charity, and inspired his neighbors to live moral and enlightened lives.

A well-run estate likewise symbolized social harmony and virtue in Jean-Jacques Rousseau's novel *Julie, or the New Héloïse*.[63] After her intense love affair with her tutor Saint-Preux, the virtuous heroine Julie married her father's friend Wolmar and moved to his estate of Clarens. Years later, Saint-Preux visited the Wolmar family. Initially wary of Monsieur de Wolmar and uncertain how to behave around Madame de Wolmar, Saint-Preux found himself welcomed into their home with much warmth. The more he saw of Clarens, the more impressed he became, and he eventually wrote a long letter to his confidant Milord Edouard in which he marveled at how well the Wolmars managed their estate. Clarens produced much, he noted, and boasted a fertile vineyard and pastures well-manured by sheep. Wolmar understood agriculture and particularly grasped the concept that the wealth of the land was near infinite as long as it was cultivated carefully and by many hands. "Monsieur de Wolmar contends that land produces in proportion to the number of hands that till it," Saint-Preux wrote. "Better tilled it yields more; this excess production furnishes the means of tilling it better still; the more men and beasts you put on it, the more surplus it supplies over and above their subsistence." Like the physiocrats, Wolmar understood nature's bounty to be plentiful and mysteriously capable of providing for those who worked the land, even if "it is not known . . . where this continual reciprocal increase in product and laborers might end."[64] A wise and attentive manager like Wolmar ensured that the land would be fertile.

More so than agriculture, the relationships between Monsieur and Madame de Wolmar and their workers astounded Saint-Preux. Clarens was an island of social harmony, divorced from the contentious relationships between master and servant, worker and employer that threatened the peace in so many villages. This harmony especially impressed in light of the high standards the Wolmars set for their servants. Monsieur de Wolmar proved a diligent supervisor, constantly spurring his employees to work harder, live virtuously, and devote themselves fully to him and his family. Saint-Preux noted approvingly that "Monsieur de Wolmar is principled and stern, and never allows the procedures of favor and benevolence to degenerate into custom and abuse. . . . Moreover, Monsieur de Wolmar checks on [the workers] himself nearly every day, often several times a day, and his wife likes to join in these rounds." Master and mistress bestowed honors on those workers who especially pleased them: "during the peak labors, Julie each week gives a gratification of twenty *batz* to the one worker of all . . . who in the master's judgment has been the most diligent during that week."[65] This contest encouraged employees to work harder and longer than they otherwise would have, ensuring that the Wolmars' profits more than made up the cost of the small prize. Profit and paternalism could go hand in hand.

The Wolmars held themselves up as models of virtue for their servants to emulate. They were in turn beloved by their employees, who recognized their loving hearts and wise management. Julie in particular had a way of connecting with her female servants: "founded on confidence and attachment, the familiarity that reigned between the servants and the mistress only strengthened respect and authority, and the services rendered and received seemed to be tokens of reciprocal friendship."[66] Julie's servants recognized her virtue and aspired to be more like her, to be friends with her. Indeed, the Wolmars relied heavily on emulation, encouraging workers to imitate the most diligent among them and to follow the virtuous example set by master and mistress.

Rousseau thus depicted the Wolmars as renewing not only their land but also the social bonds of all those around them. Julie de Wolmar took a particularly hands-on approach with her employees: "Workers, servants, all those who have served her, be it for a day, become her children. She shares in their pleasures, in their sorrows, in their lot; she inquires about their business, their interests are hers." The task of inspiring love and devotion was a full-time occupation, it would seem: "she takes on a thousand cares on their behalf, she gives them advice, she patches up their disputes, and expresses the affability of her character not with honeyed, ineffectual words, but with genuine services and continual acts of kindness." Her efforts were well-rewarded, as Julie found herself the object of her servants' admiration: "they, in return, drop everything at her slightest sign; they fly when she speaks; a mere look from her inspires their zeal, in her presence they are content, in her absence they talk about her."[67] Nothing, it would seem, was outside Julie's purview.

Masters had long had a broad mandate to dictate all manner of behavior to their servants: what to wear, where to worship, and how to behave. The Wolmars represented a change to this model, however, in that their control was accomplished through a secular paternalism. The Wolmars did not appeal to God's authority or harangue their servants with biblical phrases. Instead, Julie mothered them into obedience and loyalty; her husband fulfilled the role of stern but beneficent father.

The Wolmars were the picture of virtuous estate management: learned in agricultural techniques, beloved by their workers. This was no accident. As Sarah Maza has noted, the Wolmars skillfully manipulated their servants' time and salaries to ensure their undivided loyalty.[68] The contests, the surveillance, the loving attention: the Wolmars deployed all of these to extract the maximum amount of labor, devotion, and loyalty out of their workers. In so doing, they made Clarens efficient, orderly, fertile, and well-populated. It was the stuff agronomists' dreams were made of.

Lavoisier likewise saw his work at Fréchines as comprehensive, secular, and paternalist. Like Rousseau, he believed in the natural goodness of agriculture and peasant life; like the Wolmars, he did more than simply manage planting and harvesting. His wife's eulogy, in which she depicted him "acting as a magistrate of the peace to reestablish good relations between two neighbors, or to return a son to paternal obedience; in general, giving an example of all patriarchal virtues," even echoed Rousseau's description of Clarens: Julie herself "takes on a thousand cares on their behalf, she gives them advice, she patches up their disputes, and expresses the affability of her character not with honeyed, ineffectual words, but with genuine services and continual acts of kindness."[69] In both cases, a secondhand observer represented the protagonist as an ideal authority figure: someone devoted to caring for others and ensuring the social harmony of everyone in the vicinity.

The resemblance may have been intentional. Lavoisier, like so many eighteenth-century readers, found himself sucked into Rousseau's novel, and he ruminated on the love affair between Julie and Saint-Preux. He copied key dialogue between the young lovers and then wrote extensive descriptions setting the scenes discussed in their letters.[70] Although Lavoisier did not address the scenes at Clarens in these musings, the fact that he copied passages from the book and mulled over the physical setting for the scene suggests that he read the novel carefully, perhaps even multiple times. Many of his peers made a similar emotional investment in Rousseau's work, weeping, gasping, and sighing as they read. As such, it seems plausible that Lavoisier would have been intimately familiar with the goings-on at Clarens and may even have modeled his conduct at Fréchines after that of the Wolmars.

All these landlords, real and fictional, congratulated themselves for improving the morals of their neighbors at the same time that they amped up agricultural productivity and promoted economic opportunities. Their acts of *bienfaisance* illustrated their natural virtue and showed that they, as truly good people, took pleasure in helping others.[71] For that reason, they believed themselves to be good patriots as well as good landlords. This idea of patriotism neatly fit with Enlightenment ideas of secular morality and theories of society. Good patriots were motivated by love of humanity and in particular their fellow citizens. This was a very different economic world from the one we inhabit today. Unlike our hard-boiled vision of capitalism, eighteenth-century economic life was infused with sentiment and social connection. Thinkers drew on "cold" calculation as well as "warm" emotions, neatly weaving together personal and economic interests.[72] The ideal landlord did not see economic calculation and sentiment as an either-or proposition; instead, he or she should aspire to boost the estate's productivity and to form sentimental bonds with neighbors and workers. Sweet sentiments and hard efficiency went together, at least in theory.

These ideas were central to the way Lavoisier understood his work on his estates. His work not only benefited humanity because of the knowledge he generated; it benefited the estate's residents in particular because they were the recipients of Lavoisier's *bienfaisance*. By acting as he did, Lavoisier showed himself to be dedicated not only to the discovery of empirical truths but also to the improvement of society and humanity's lot in life. He made clear in his representations of agricultural work that "I can only have been animated . . . by my zeal for the public good."[73]

Which brings us back to his call to arms, in which he urged his fellow philosophes to leave their studies (*cabinets*). Lavoisier believed himself to be doing exactly that when he bought land and turned it into an experimental farm. In stressing that philosophes had to leave their offices occasionally, Lavoisier portrayed the ideal philosophe as engaged with the public and doing research out in the world. In keeping with this model, his agricultural endeavors were not solely focused on the collection of data but also toward modeling the ideal agricultural reformer. He represented himself as a loving and benevolent landlord as well as a learned agronomist and effective government servant. Fréchines was much more than a data set to Lavoisier. He did not just use the estate to generate agricultural statistics but also used it as a flattering backdrop for his experiments in self-fashioning.

## WHO DID THE WORK?

For all that Lavoisier stressed his personal experience with agriculture, however, he depended on his managers to handle everyday business, and he could not supervise them as closely as Wolmar did in *La Nouvelle Héloïse*. He visited Fréchines for harvesting and planting and, generally speaking, did not spend long stints on the farm. Lavoisier's many engagements often kept him occupied in Paris. At this point, he was a member of the General Farm, an active member of the Académie des Sciences, and a member of the Gunpowder Commission. His correspondence shows him to have kept a busy schedule. He crammed his scientific investigations into the early hours of the morning and the late hours of the night.[74] It was accordingly difficult to find long stretches of time to spend at Fréchines, and Lavoisier lamented that "it is rare that I can make more than three trips per year [to the estate], and these trips are only two or three weeks long. I choose to visit, as much as it is possible for me to do so, during planting and harvest."[75] Because Lavoisier had to divide his time between so many occupations, he could not permanently reside at Fréchines.

Lavoisier solved this problem by hiring an agent, Louis-Michel Lefebvre, to serve as his representative on the estate. In his report to the Société d'Agriculture, Lavoisier stressed his total confidence in Lefebvre's capabilities: he was "a person of great precision, who lives in Blois, and who never goes more than fifteen days without visiting my holdings." Moreover, Lavoisier continued, the society could share his faith in Lefebvre on account of his family connections: "I will greatly add to the confidence of the Society in telling you that the person conducting these observations is the brother of M. l'abbé Lefèvre, our colleague; he has the same zeal and energy as his brother and is highly intelligent."[76]

Lefebvre did indeed keep in close contact with the Lavoisiers. On this matter as in so many others, Marie-Anne Lavoisier assisted her husband with his correspondence. Lefebvre even kept her abreast of his personal news, on one occasion noting with pride that "Madame Lefebvre . . . gave me the gift of a son yesterday." He hastily assured her, however, that although this happy event had forced him to miss an appointment at Fréchines, he planned to return to the estate forthwith. "I will give your interest all our attention despite the small inconvenience that has appeared," he wrote.[77] Later, during the Revolution, he confirmed that all remained well on the estate: "There is nothing new happening here; everything is as tranquil as possible."[78]

Lefebvre managed the estate; the actual work fell to laborers. Although such people were crucial to the progress of agronomy in France, they rarely appeared in either agronomical or physiocratic texts. Indeed, the *Encyclopédie* included no entry for "peasant," even though it featured lengthy articles on the need for agricultural improvement. Agricultural improvers saw peasants as impediments to reform, not the instruments by which it would be accomplished.[79] The article on the "laboureur" (farm laborer) was quite brief, especially in comparison with the much lengthier articles on "fermiers" (farmers) and "grains," both of which involved substantial explications of physiocratic philosophy written by Quesnay.[80]

Like the encyclopédistes, Lavoisier himself had little to say about farm laborers. He did make observations about manual labor when a neighbor of his imported fifty Indian weavers to France to work in his linen factory, and these observations provide insight into how he assessed manual workers. Lavoisier found the Indians fascinating and—with his wife as a coauthor—recorded his observations of these workers and their families. In a manuscript they titled "Observations on Indian Families," they esteemed the Indian workers as diligent and focused on their work.[81] The Indians ate and dressed simply, and they had a quiet virtue about them. They had strong family ties, and the Lavoisiers were particularly impressed that even unmarried women conducted themselves chastely.

In short, these were exemplary laborers in every respect. They were industrious, virtuous, and communally minded. Even if the Lavoisiers did not explicitly consider the Indians a model for how workers at Fréchines should behave, the text nevertheless offers insights into what sorts of categories they would assess workers by and what sorts of behavior they hoped to see. They were not simply focused on the product itself, but also in the virtue and morality of the producers. The categories by which they assessed Indian workers—their clothing, moral conduct, work ethic, and talent—echoed the descriptions of workers in *La Nouvelle Héloïse*. The sentimental ideal of the estate was apparently one that the Lavoisiers found truly compelling.

The absence of workers from most descriptions of Fréchines suggests that neither Antoine nor Marie-Anne Lavoisier considered them significant actors. Rather, they were acted on. They were agriculture's version of "invisible technicians": they provided manual labor, but were not involved in designing or contemplating reform in any real way.[82] Although the workers themselves surely felt differently, the Lavoisiers considered laborers to be pawns in their plans for reform—potentially virtuous, pitiable pawns, but pawns nonetheless.

## LOANS AND LIBERTY TREES

Although Lefebvre assured the Lavoisiers in 1790 that "there is nothing new happening here," the Revolution led Lavoisier to change strategies at Fréchines. He increasingly used the farm as proof of his patriotism and political service. During the poor harvest of 1788, when crops throughout France faltered due to poor weather conditions, Lavoisier worked to protect his adopted region of Blois. To moderate the effects of the crisis in his region, Lavoisier offered the Blois assembly "a sum of 50,000 *l.* to be used to purchase wheat from neighboring provinces, which seemed sufficient to the king's procurer to guarantee the subsistence of this province; I offer you this sum as a loan without interest through next September."[83] This interest-free loan would help prevent starvation and utter ruin. He also loaned 6,000 *livres*, albeit anonymously, to the nearby town of Romorantin.

Lavoisier's efforts to alleviate the crisis were in keeping with his general philosophy as well as his prior efforts to soften the effects of poor harvests. Given the broader political context, however, Lavoisier may have acted with an eye to his future. He aspired to represent Blois in the nascent Estates General, although late eighteenth-century political culture discouraged him or any other candidate from campaigning for office. Instead, the electors were meant to discern which individuals were the most virtuous and able to represent the interests of the nation. To campaign would be arrogant, ambitious, and selfish and would disqualify a candidate from further consideration.[84] Ideal politicians would be virtuous and unassuming, qualities that ambition would undermine. Lavoisier had spent years cultivating a reputation as a man of virtue and had worked especially hard to portray himself as a good citizen of Blois, but he had also been lampooned for his unsavory association with the General Farm and for the vast sums he had earned through his employment there. Lavoisier had made his fortune as a tax farmer, and tax farmers were hardly popular figures. Perhaps Lavoisier hoped his generous loans would muffle criticism and help him earn the trust and the votes necessary to secure the political position he desired. Perhaps, too, he imagined that the loans would remind electors of his comfortable financial status and reassure them that he would be able to make disinterested, impartial decisions on behalf of the public good, unlike those who were more dependent on the salaries or largesse of others.[85]

The reaction to Lavoisier's generosity—politically motivated or not—shows that he read the situation well. Upon learning of Lavoisier's actions, the Assemblée des Messieurs du Clwb de Blois wrote that they wished to give "a sign of their gratitude for the obliging manner with which he [Lavoisier] had contributed to cast Enlightenment here and into the affairs which currently affect the Nation, and for the kindness of his actions toward our province in the afflicting circumstances in which she finds herself."[86] Lavoisier's response to this praise was steeped in humility and patriotism. As he wrote to the municipal officers of Blois, "Nothing could flatter me more than the honorable title of Citizen of Blois which you have conferred on me, because this is an expression of sentiment. It is a new line that attaches me to your province and to your town. Blois has become my adoptive homeland, and it is no less dear to me than the place where I was born." He wished to make clear that more than agricultural profits or scientific opportunity linked him to Blois; in his heart, he felt connected to the region and to his fellow citizens. He swore "an oath of fidelity that I make to you in accepting the title of fellow citizen. Your interests will always be dear to me, and I will always make it my duty to support you if circumstances should ever call me to defend you."[87]

But Lavoisier soon discovered that he had to find a way to tactfully decline making new loans. Tales of his open coffers soon spread beyond Blois and Romorantin, and he received a letter from the representative of Vendôme soliciting assistance. He wrote back that, regretfully, "I have exhausted myself for months to prevent the awful misfortunes which have menaced the towns of Blois and Romorantin. . . . I thus find myself obliged to beg your permission, Messieurs, to let me contribute only 300 *livres* toward the loss you have suffered. In happier times I would have offered you more."[88] Although Lavoisier was wealthy, his personal fortune could not make up for the losses suffered by every village. He concentrated his efforts on assisting the towns closest to Fréchines, testimony to how closely he identified his interests and his patriotism with his estate.

Lavoisier's dreams of representing his fellow citizens in the National Assembly ended when he failed to be elected. Even after the disappointing results of the Estates General elections, however, Lavoisier continued to provide moral and financial support to his fellow Blois patriots. In 1790, when the town planned to send a regiment to Paris for the Festival of the Federation, he wrote "in my capacity as citizen of Blois and supplementary deputy for that same city in the National Assembly, I feel obliged to offer lodging in Paris to the representatives of the department of Blois."[89] The deputies had not yet been named but Lavoisier hoped that, whoever they might be, they would accept his offer. He soon discovered—happily so— that his representative at Fréchines, Lefebvre, was one of those chosen.

As Lavoisier's political role evolved, however, one constant was that he used Fréchines as a backdrop for his patriotism. He listed Blois, not his birthplace and longtime hometown, Paris, as his official residence. He ceremoniously planted "liberty trees" in Blois during the Revolution. That he represented himself as a citizen of Blois suggests that he considered his life at Fréchines to provide significant evidence of his commitment to his nation and to the ideals of the Revolution. The estate continued to be a site on which Lavoisier displayed an idealized version of himself: a revolutionary patriot.

## CONCLUSION

Lavoisier's work at Fréchines provides a window into philosophes' sense of themselves and their role in the world on the eve of the Revolution. Lavoisier and many of his fellow savants believed that their purpose was to reshape society, sometimes in fundamental ways, and to do so by working in the world as well as in their studies. The history of Enlightenment ideas of social reform tends to come across as abstract. Yet on a microscale—in their homes, on their estates—savants like Lavoisier tested their ideas and modeled their successes.

In analyzing such practices, this chapter has built on the arguments advanced in previous chapters to show that a paternal style of thinking operated outside the family home. Lavoisier's dealings at Fréchines were paternalist, even if he was no one's *pater*. Because eighteenth-century individuals saw the family as a model for society, family and fatherly language easily applied to circumstances outside the family home. Although an experimental farm might seem to have little to do with marrying one's beloved, working alongside one's relatives, or inoculating and educating one's children, Lavoisier's work at Fréchines fit within the same mode of intellectual and social engagement.

At the same time, Lavoisier held himself up as a practitioner rather than an advisor. It was not enough for him to read up on agronomy, physiocracy, liberalism, and sentimental estate management; he had to appear to live those ideals. In this way, he established his scientific and moral authority. His actions at Fréchines were, at every step, particular rather than abstract. He highlighted his emotional connection to the area and its people, presenting himself as warm and attached rather than rational and detached. Although the estate was only one of his homes, Lavoisier found Fréchines a particularly useful site for self-fashioning. Showing himself immersed in agricultural reform, committed to economic renewal, and dedicated to the well-being of his neighbors enabled Lavoisier to portray himself as patriotic, charitable, and devoted to the public good.

In sum, Lavoisier used Fréchines to present his ideas in action and to cast the best possible light on his improvements. He represented himself as a diligent savant dedicated to capturing the effects of agronomical reform in minute detail, a paternalistic landlord who expended considerable time and energy caring for his neighbors, and a patriotic citizen willing to make personal sacrifices to help those less fortunate than he. This was not some arcadian idyll where he could escape the world: Lavoisier drew the attention of his peers to his actions at Fréchines. That his memorialists continued to do the same after his untimely death is also suggestive. For them, as for Lavoisier himself, it was important to note for posterity that he was not just a brilliant chemist. He was also a man whose death caused peasants to weep.

# Conclusion

When Condorcet wrote that "one finds, among bachelors and married men, men of equal genius," he sought to bring a longstanding debate to a close.[1] Married men of letters need not be ashamed that they had fallen in love and married; they need not duck their heads as "the husband ashamed to be one," like the protagonist of Destouches's 1727 play *Le Philosophe Marié*. They should instead take pride in their wife and children and cherish the love that they (ideally) shared. Bachelor life had its advantages, and Condorcet was quick to note that there was no need to chastise philosophers who remained single. But married life no longer seemed quite like a ball and chain. Philosophes who embraced family life insisted that domestic life made them happier, more productive, and more virtuous. They were better suited to advise the public on a host of matters because of, not in spite of, their blissful home lives. By the middle of the eighteenth century, the language of sentimental domesticity had become more important than ever before in proving one's sociability, public utility, and sensitivity. Being good fathers and good husbands helped thinkers seem like ideal public men and ideal men of letters. The importance of these qualities would be born out during the French Revolution, when many citizens depicted themselves as loving family men and dedicated citizens of the *patrie*.

The stoic, unmarried philosopher remained a powerful ideal during and long after the Enlightenment, but married men of letters discovered they had a bag of tricks unavailable to the unattached. Many philosophers were eager to step into public roles and to reach wide audiences with their works. Married men of letters drew attention to their loving families as proof that they were men of virtue more than qualified to make themselves useful to the public, to stand as role models as well as men of learning and wit. They called for the public to imitate them in their roles as fathers, to inoculate and educate their children just as philosophes had inoculated and educated their own children. Women of letters used the same kind of rhetoric, wrapping their writings on natural philosophy, education, and inoculation in a warm blanket of maternal solicitude. Men and women embraced an intimate empiricism, attempting to live their lives according to Enlightenment ideals and using personal experiences to inject their correspondence and published writings with greater verisimilitude, compelling empirical evidence, and sentiment.

In myriad ways, the family served as an institution of the Enlightenment. Wives and children performed all manner of useful tasks ranging from domestic maintenance to the observation and calculation of data. It was once commonplace to speak of Enlightenment philosophes as geniuses who worked alone on their texts, but studying the making of Enlightenment knowledge from the angle of the family workshop underscores yet again the fundamentally collective ways in which ideas were developed and advertised. Wives, daughters, sons, and relatives were active participants in many *cabinets*, laboratories, and observatories. Such work acquired a positive moral valence in the eighteenth century, the age of sentiment. Although women savants remained controversial figures, vulnerable to any number of character attacks, women working with their families enjoyed a certain amount of protection. They and their male family members developed a newly emotional way of discussing family labor by stating that these women demonstrated their love for their savant husbands or fathers through their work.

Most fundamentally, the family home acted as a backdrop for savants eager to forge new public personae. Representing themselves as loving fathers and mothers allowed savants to tap into a new moral authority, to represent themselves as public figures worthy of emulation. Much work on early modern intellectual self-fashioning has focused on laboratories and academies, but *Sentimental Savants* has argued that more intimate contexts mattered as well. Many savants represented themselves as cozy domestic figures as well as masters of empirical observation and well-connected academicians. The sentimental savant was another facet of self-fashioning during the Enlightenment, and it coexisted easily with other, more familiar models of intellectual authority.

Viewing savants through the lens of intimate life makes them seem less like abstract, impractical thinkers and more like adaptable and pragmatic figures interested in applying their ideas to their own lives. In many respects, the kinds of social experiments I have discussed here were the first iteration of efforts to reform "the social," a category that only grew in analytical and political power as the eighteenth century waned. Indeed, during the French Revolution, those who had been weaned in the Age of Enlightenment expressed great confidence in their abilities to remake their world as they saw fit. They jettisoned long-held traditions and imagined a bold new world full of possibilities, possibilities that included sweeping political, social, and cultural reforms.

In drawing attention to their intimate lives, Enlightenment savants initiated a dance between personal and intellectual, public and private, that waltzed on long after they had left the floor. When revolutionaries insisted that only married priests were true patriots, and when the Jacobin Camille Desmoulins posed for a portrait with his patriotic family rather than sitting alone, they danced to the same tune.[2] When Charles Darwin monitored his children's emotions and deputized his daughter and son-in-law (on their honeymoon, no less) to collect botanical specimens from far-off locales, he did the same.[3] Recent research suggests that there is still much to learn about the personal foundations of public virtue and intellectual authority, but it is abundantly clear that the back-and-forth between intimate and intellectual, private and public did not draw to a close with the eighteenth century.

Paternal and political authority sometimes rested uneasily on claims of personal virtue, but the fact that scientists, philosophers, and politicians kept coming back to the well of family love suggests that domestic sentiments remained a powerful component of public personae. To be sure, other representations of public and intellectual life abounded: the scientist working with male students in his laboratory; the rise of new, more powerful academic institutions, universities, and museums; a decorous divide between a politician's public deeds and private actions. But the separation between these images and those set within family homes was not absolute. Scientific laboratories could function as a sort of family; wives acted as power brokers for men of letters and politicians in the post-revolutionary era.[4] The relationship between families, philosophy, science, and the public has ebbed and flowed but never totally receded.

ACKNOWLEDGMENTS

I have looked forward to writing this page for some time, both because it is a pleasure to acknowledge those who have helped me with this project and because writing the acknowledgments means I am almost done with this book! I am very grateful for the material assistance I have received in support of this project: a Jacob K. Javits fellowship from the US Department of Education, a Millstone Fellowship from the Western Society for French History, a Presidential Fellowship from Northwestern University, and an Andrew W. Mellon Fellowship from Bowdoin College. Various grants from Bowdoin College have also funded my research.

I must salute the teachers and mentors who have shaped this project in ways big and small. Sarah Maza has been an outstanding mentor. Her comments are always incisive and she is great fun to boot. I am profoundly grateful to her. Ken Alder has been an excellent guide to the history of science and his thoughts on my manuscript improved it in myriad ways. Ed Muir's enthusiasm for early modern history helped spark and sustain my own interests. Ron Schechter's course on the age of absolutism introduced me, many years ago, to the history of eighteenth-century France. LuAnn Homza is always willing to talk through my ideas and to offer sound counsel. And I probably wouldn't be a historian at all if it weren't for Scovie Martin, my AP US history teacher at Western Albemarle High School, who helped me develop my love for historical research.

I presented facets of this work at various conferences and am especially grateful to my colleagues in the Western Society for French History, the Society for French Historical Studies, and the History of Science Society. The questions I received at these conferences helped me develop my ideas, and I am also grateful for the scholarly camaraderie facilitated by these meetings. I benefited a great deal from a seminar on family values hosted by Warwick-in-Venice. Thanks also to the many colleagues who provided valuable feedback on my work: Jeremy Caradonna, Denise Davidson, Nina Gelbart, Jan Golinski, Dena Goodman, Jennifer Heuer, Jeff Horn, Nina Kushner, Elise Lipkowitz, Carol Pal, Jennifer Popiel, Jessica Riskin, Sarah Ross, April Shelford, Mary Terrall, and Kathleen Wellman.

I've made many new friends while working on this project. I have to agree with Sarah Vowell, who once wrote: "Being a nerd, which is to say going too far and caring too much about a subject, is the best way to make friends I know." Stefanie Bator, Charlotte Cahill, and Celeste McNamara offered sage advice and good times over glasses of beer; my fellow Europeanists Will Cavert, Genevieve Carlton, and Jason Johnson made for excellent writing buddies at Northwestern; Natasha Naujoks, Mary Elizabeth O'Neill, and Amy McKnight helped make my research trips to Paris something to look forward to; Margaret Boyle, Kelly Fayard, Peggy Wang, Greg Beckett, and Ingrid Nelson have been great colleagues and writing partners at Bowdoin. Dallas Denery, Arielle Saiber, Ann Kibbie, Aaron Kitch, and Robert Morrison read drafts of my work and make Bowdoin a fun place to be a premodernist. My department colleagues have been supportive of my work, and Page Herrlinger deserves particular thanks for being an exemplary chair.

An earlier version of chapter 1 appeared as "*Philosophes Mariés* and *Epouses Philosophiques*: Men of Letters and Marriage in Eighteenth-Century France," *French Historical Studies* 35, no. 3 (Summer 2012), 509–39; and a section of chapter 2 is forthcoming in the *Journal of Women's History* as "Learned and Loving: Representing Women Astronomers in Eighteenth-Century France." I am grateful to the editors and publishers of these journals for permission to reprint.

My book team at the University of Chicago Press has been a joy to work with. Karen Darling expertly guided this book through the acquisitions process and has been a wonderfully thoughtful editor. Evan White has ably fielded my questions about formatting, permissions, and the like. Kelly Finefrock-Creed meticulously edited the final text. Many thanks are due to my two anonymous reviewers, who were models of what academic peer review can be: learned, thoughtful, and encouraging.

I owe my family a great deal, even if they're still a bit perplexed why I have devoted my career to the study of French history. My mother, Kathleen Garcia, sacrificed a great deal to ensure that I could pursue a top-flight education. My grandfather, Ciro Robustelli, insisted that I should apply myself to any task at hand. My sister, Olivia Garcia, makes me laugh and my brother, Brendan Cunningham, always impresses me with his grasp of historical thinking.

Last but most certainly not least, I must thank my spouse and son. Ciro Roberts was born the same week that I learned this book was accepted for publication, and he brought much joy to the final stages of work. Strother Roberts has been there since the very beginning. He read the entire manuscript multiple times and is an unflagging source of support. He deserves effusive praise in the style of the age of sentiment ("Oh most worthy spouse!" etc.) but, as a practical Midwesterner, would undoubtedly appreciate something simpler. Strother, thank you. This book is for you.

INTRODUCTION

1. Many works touch on these subjects, but see especially Sarah Hanley, "Engendering the State: Family Formation and State Building in Early Modern France," *French Historical Studies* 16, no. 1 (Spring 1989): 4–27; Jeffrey Merrick, "Fathers and Kings: Patriarchalism and Absolutism in Eighteenth-Century French Politics," *Studies on Voltaire and the Eighteenth Century* 308 (1993): 281–303; Sarah Maza, *Private Lives and Public Affairs: The Causes Célèbres of Pre-revolutionary France* (Berkeley: University of California Press, 1993); David Denby, *Sentimental Narrative and the Social Order in France, 1760–1820* (Cambridge: Cambridge University Press, 1994); Lynn Hunt, *The Family Romance of the French Revolution* (Berkeley: University of California Press, 1993); Suzanne Desan, *The Family on Trial in Revolutionary France* (Berkeley: University of California Press, 2004).

2. On the incompatibility of family life and philosophy, see Robert Darnton, *The Literary Underground of the Old Regime* (Cambridge, MA: Harvard University Press, 1982), 14–15; Rémy Sausselin, *The Literary Enterprise in Eighteenth-Century France* (Detroit, MI: Wayne State University Press, 1979), 24. On married men of letters, see Anne C. Vila, "Faux Savants, Femmes Philosophes & Philosophes Amoureux: Foibles of the Philosophes on the Eighteenth-Century French Stage," *Studies in Eighteenth-Century Culture* 35 (2006): 203–21. On philosophical identity in general, see Dinah Ribard, *Vivre, Raconter Penser: Histoire(s) des Philosophes, 1650–1766* (Paris: Éditions EHESS, 2003).

3. For example, Jean-Pierre Poirier, *Lavoisier: Chemist, Biologist, Economist*, trans. Rebecca Balinski (Philadelphia: University of Pennsylvania Press, 1996); Judith P. Zinsser, *La Dame d'Esprit: A Biography of the Marquise du Châtelet* (New York: Viking, 2006); Carol Blum, *Diderot: Virtue of a Philosopher* (New York: Viking, 1974).

4. I stop short, however, of labeling savants or sentimental domesticity as bourgeois. This reluctance draws on recent scholarship on eighteenth-century France, which has illustrated that there was no coherent middle-class consciousness at the time. See Sarah Maza, *The Myth of the French Bourgeoisie: An Essay on the Social Imaginary, 1750–1850* (Cambridge, MA: Harvard University Press, 2003). My reluctance also stems

from prior work on the philosophes as family men, most notably an older study by Robert Darnton which overattributed philosophes' interest in family life to their status as men of the middle class. Darnton, *Literary Underground of the Old Regime*, 14–15.

5. Peter Abelard, *Story of Abelard's Adversities*, ed. and trans. J. T. Muckle (Toronto: Pontifical Institute of Mediaeval Studies, 1964), 33.

6. Elena Russo, *Styles of Enlightenment: Taste, Politics, and Authorship in Eighteenth-Century France* (Baltimore, MD: Johns Hopkins University Press, 2007).

7. On new conceptions of society, see Laurence Kauffmann and Jacques Guil-haumou, eds., *L'invention de la société: Nominalisme politique et science sociale au XVIIIe siècle* (Paris: Éditions de l'École des Hautes Études en Sciences Sociales, 2003); Daniel Gordon, *Citizens without Sovereignty: Equality and Sociability in French Thought, 1670–1789* (Princeton, NJ: Princeton University Press, 1996). On malleability and its limits, see Michael E. Winston, *From Perfectibility to Perversion: Meliorism in Eighteenth-Century France* (New York: Peter Lang, 2005).

8. Jessica Riskin, *Science in the Age of Sensibility: The Sentimental Empiricists of the French Enlightenment* (Chicago: University of Chicago Press, 2002); Anne C. Vila, *Enlightenment and Pathology: Sensibility in the Literature and Medicine of Eighteenth-Century France* (Baltimore, MD: Johns Hopkins University Press, 1998); William Reddy, *The Navigation of Feeling: A Framework of the History of Emotions* (Cambridge: Cambridge University Press, 2001), 173–210; John Mullan, *Sentiment and Sociability: The Language of Feeling in the Eighteenth Century* (New York: Oxford University Press, 1990); Denby, *Sentimental Narrative*; Lynn Hunt, *Inventing Human Rights: A History* (New York: W. W. Norton, 2007).

9. Anne Vincent-Buffault, *The History of Tears: Sensibility and Sentimentality in France*, trans. Teresa Bridgeman (New York: St. Martin's Press, 1991), 53.

10. Ernst Cassirer, *The Philosophy of the Enlightenment* (Princeton, NJ: Princeton University Press, 2009); Peter Gay, *The Enlightenment: An Interpretation; The Rise of Modern Paganism* (New York: Norton, 1966); Jonathan Israel, *Radical Enlightenment: Philosophy and the Making of Modernity, 1650–1750* (New York: Oxford University Press, 2001); Jonathan Israel, *Enlightenment Contested: Philosophy, Modernity, and the Emancipation of Man, 1670–1752* (Oxford: Oxford University Press, 2006).

11. Darrin McMahon, *Enemies of the Enlightenment: The French Counter-Enlightenment and the Making of Modernity* (New York: Oxford University Press, 2001), 56–62; Roger Chartier, *The Cultural Origins of the French Revolution*, trans. Lydia G. Cochrane (Durham, NC: Duke University Press, 1991), 4–7.

12. Robert Darnton's work on Grub Street and Dena Goodman's work on salons has been especially influential. Darnton, *Literary Underground of the Old Regime*; Dena Goodman, *The Republic of Letters: A Cultural History of the French Enlightenment* (Ithaca, NY: Cornell University Press, 1994). Recent work continues to widen our understanding of Enlightenment participation: Sean Takats, *The Expert Cook in Enlightenment France* (Baltimore, MD: Johns Hopkins University Press, 2011); Jeremy Caradonna, *The Enlightenment in Practice: Academic Prize Contests and Intellectual Culture in France, 1670–1794* (Ithaca, NY: Cornell University Press, 2012).

13. Goodman, *Republic of Letters*; Caradonna, *Enlightenment in Practice*; Eliza-beth Andrews Bond, "Letters to the Editor in Eighteenth-Century France: An Enlight-

enment Information Network, 1770–1791" (PhD diss., University of California-Irvine, 2014); Kenneth Loiselle, *Brotherly Love: Freemasonry and the Transformation of Male Friendship in Eighteenth-Century France* (Ithaca, NY: Cornell University Press, 2014); E. C. Spary, *Eating the Enlightenment: Food and the Sciences in Paris, 1670–1760* (Chicago: University of Chicago Press, 2012).

14. A phrase I borrow from Margaret C. Jacob, *Living the Enlightenment: Freemasonry and Politics in Eighteenth-Century Europe* (Oxford: Oxford University Press, 1991).

15. Lynn Hunt, *Politics, Culture, and Class in the French Revolution* (Berkeley: University of California Press, 1986); Desan, *Family on Trial*; David Bell, *The Cult of the Nation in France: Inventing Nationalism, 1680–1800* (Cambridge, MA: Harvard University Press, 2003); Sophia Rosenfeld, *Revolution in Language: The Problem of Signs in Late 18th-Century France* (Palo Alto, CA: Stanford University Press, 2001).

16. William Reddy, *The Navigation of Feeling: A Framework of the History of Emotions* (Cambridge: Cambridge University Press, 2001), 129.

17. Steven Shapin and Simon Schaffer, *Leviathan and the Air Pump: Hobbes, Boyle and the Experimental Life* (Princeton, NJ: Princeton University Press, 1989).

18. Riskin, *Science in the Age of Sensibility.*

19. Pamela Smith, *Body of the Artisan: Art and Experience in the Scientific Revolution* (Chicago: University of Chicago Press, 2004); Deborah Harkness, *The Jewel House: Elizabethan London and the Scientific Revolution* (New Haven, CT: Yale University Press, 2007).

20. Londa Schiebinger, *The Mind Has No Sex? Women in the Origins of Modern Science* (Cambridge, MA: Harvard University Press, 1991); Patricia Fara, *Pandora's Breeches: Women, Science, and Power in the Enlightenment* (London: Pimlico, 2004); Pnina G. Abir-Am and Dorinda Outram, eds., *Uneasy Careers and Intimate Lives: Women in Science, 1789–1979* (Newark, NJ: Rutgers University Press, 1987).

21. Paul White, "Darwin's Emotions: The Scientific Self and the Sentiment of Objectivity," *Isis* 100, no. 4 (December 2009): 811–26; Sarah Ross, *The Birth of Feminism: Woman as Intellect in Renaissance Italy and England* (Cambridge, MA: Harvard University Press, 2009); Deborah Coen, *Vienna in the Age of Uncertainty: Science, Liberalism, and Private Life* (Chicago: University of Chicago Press, 2007); John Randolf, *The House in the Garden: The Bakunin Family and the Romance of Russian Idealism* (Ithaca, NY: Cornell University Press, 2007). For related work, see Emily J. Levine, "PanDora, or Erwin and Dora Panofsky and the Private History of Ideas," *Journal of Modern History* 83, no. 4 (December 2011): 753–87, 757; Helena M. Pycior, Nancy G. Slack, and Pnina Abir-Am, eds., *Creative Couples in the Sciences* (New Brunswick, NJ: Rutgers University Press, 1996); *For Better or For Worse? Collaborative Couples in the Sciences*, ed. Annette Lykknes, Donald Opitz, and Brigitte Van Tiggelen (Boston: Birkhauser, 2012).

22. A notable exception is Mary Terrall, *Catching Nature in the Act: Réaumur and the Practice of Natural History in the Eighteenth Century* (Chicago: University of Chicago Press, 2014).

23. Barbara H. Rosenwein, "Problems and Methods in the History of Emotions," *Passions in Context* 1 (2010): 11; Barbara Rosenwein, *Emotional Communities in the Early Middle Ages* (Ithaca, NY: Cornell University Press, 2006); for a general discus-

sion of the history of emotions, see "The Historical Study of Emotions," a roundtable in *American Historical Review* 117, no. 5 (December 2012): 1487–1531.

24. The transnational case is made forcefully in Jonathan Israel, *Radical Enlightenment: Philosophy and the Making of Modernity, 1650–1750* (New York: Oxford University Press, 2001); Charles W. J. Withers, *Placing the Enlightenment: Thinking Geographically in the Age of Reason* (Chicago: University of Chicago Press, 2007); Neil Safier, *Measuring the New World: Enlightenment Science and South America* (Chicago: University of Chicago Press, 2008).

CHAPTER ONE

1. Jean-Jacques Rousseau to Suzanne Dupin de Francueil, 20 April 1751, in *Correspondance*, ed. R. A. Leigh (Geneva: Institute et Musée Voltaire, 1965), 2:142. Unless otherwise noted, all translations are my own.

2. Pierre Choderlos de Laclos to Madame Laclos, 18 floréal an 2, in *Lettres inédites de Choderlos de Laclos*, ed. Louis de Chauvigny (Paris: Mercure de France, 1904), 1:38. For thoughts on why Laclos, author of the libertine novel *Les Liaisons Dangereuses*, wrote such uncynical love letters to his wife, see Philippe Sollers, "What Is Libertinage?" *Yale French Studies* 94 (1998): 204–5.

3. Jérôme Lalande, "Testament Moral," in Louis Aimable, *Le Franc-Maçon Jérôme Lalande* (Paris: Charavay Frères, 1889), Appendix M, 52.

4. On financial constraints as a potential obstacle to marriage, see Robert Darnton, "A Police Inspector Sorts His Files: The Anatomy of the Republic of Letters," in *The Great Cat Massacre and Other Episodes in French Cultural History* (New York: Vintage Books, 1985), 169–72; David J. Sturdy, *Science and Social Status: The Members of the Académie des Sciences, 1666–1750* (Woodbridge, UK: Boydell Press), 267; Bruno Belhoste, *Paris Savant: Parcours et rencontres au temps des Lumières* (Paris: Armand Colin, 2011), 15.

5. For more on the lack of a self-conscious bourgeoisie in France, see Sarah Maza, *The Myth of the French Bourgeoisie: An Essay on the Social Imaginary, 1750–1850* (Cambridge, MA: Harvard University Press, 2003), chaps. 1–2.

6. Anne C. Vila, "Faux Savants, Femmes Philosophes & Philosophes Amoureux: Foibles of the Philosophes on the Eighteenth-Century French Stage," *Studies in Eighteenth-Century Culture* 35 (2006): 203–21; Dinah Ribard, *Raconter, Vivre, Penser: Histoire(s) de philosophes, 1650–1766* (Paris: Vrin, 2003), 113–36.

7. Gadi Algazi, "Scholars in Households: Refiguring the Learned Habitus, 1480–1550," *Science in Context* 16, no. 1 (2003): 9–42; Ruth Mazo Karras, "Sharing Wine, Women, and Song: Masculine Identity Formation in the Medieval European Universities," in *Becoming Male in the Middle Ages*, ed. Jeffrey Jerome Cohen and Bonnie Wheeler (New York: Garland Publishing, 1997), 192–95.

8. Algazi, "Scholars in Households," 18.

9. Anthony F. D'Elia, *The Renaissance of Marriage in Fifteenth-Century Italy* (Cambridge, MA: Harvard University Press, 2004), 117–34; Charles Taylor, *Sources of the Self: The Making of Modern Identity* (Cambridge, MA: Harvard University Press, 1992), 213; Paul Oskar Kristeller, "The Active and the Contemplative Life in Renais-

sance Humanism," in *Arbeit, Musse, Meditation: Betrachtungen zur Vita Activa und Vita Contemplativa*, ed. Brian Vickers (Zurich: Verlag der Fachvereine, 1985), 142.

10. Daniel Gordon, *Citizens without Sovereignty: Equality and Sociability in French Thought, 1670–1789* (Princeton, NJ: Princeton University Press, 1994), 51–85; Keith Baker, "Enlightenment and the Institution of Society: Notes for a Conceptual History," in *Main Trends in Cultural History: Ten Essays*, ed. William Melching and Wyger Velema (Amsterdam: Rodopi, 1994), 95–120; Maza, *Myth of the French Bourgeoisie*, chap. 1; for the general push toward secularization, see Daniel Roche, *France in the Enlightenment*, trans. Arthur Goldhammer (Cambridge, MA: Harvard University Press, 1998), 519–47.

11. Gordon, *Citizens without Sovereignty*, 85.

12. Anthony Ashley Cooper, 3rd Earl of Shaftesbury, *An Inquiry Concerning Virtue, in Two Discourses*, ed. Joseph Filonowicz (Delmar, NY: Scholars' Facsimiles & Reprints, 1991), 115.

13. Gordon B. Walters Jr., *The Significance of Diderot's "Essai sur le mérite et la vertu"* (Chapel Hill: University of North Carolina Press, 1971), 92–96.

14. César Chesneau Du Marsais, "Philosopher," in *The Encyclopedia of Diderot & d'Alembert Collaborative Translation Project*, trans Dena Goodman (Ann Arbor: Michigan Publishing, University of Michigan Library, 2002), http://hdl.handle.net/2027/spo.did2222.0000.001 (accessed 18 July 2014). Originally published as "Philosophe," in *Encyclopédie ou Dictionnaire raisonné des sciences, des arts et des métiers* (Paris, 1765), 12:509–11.

15. Dena Goodman, *The Republic of Letters: A Cultural History of the French Enlightenment* (Ithaca, NY: Cornell University Press, 1994); April Shelford, *Transforming the Republic of Letters: Pierre-Daniel Huet and European Intellectual Life, 1650–1720* (Rochester, NY: Rochester University Press, 2007); Carol Pal, *Republic of Women: Rethinking the Republic of Letters in the Seventeenth Century* (Cambridge: Cambridge University Press, 2012); Sylvana Tomaselli, "The Enlightenment Debate on Women," *History Workshop Journal* 20 (1985): 101–24.

16. Taylor, *Sources of the Self*, 213; *Pétition adressée a l'Assemblée Nationale* (Paris: Imprimerie Cercle Social, 1790); Suzanne Desan, *The Family on Trial in Revolutionary France* (Berkeley: University of California Press, 2004), 47–92; James F. Traer, *Marriage and the Family in Eighteenth-Century France* (Ithaca, NY: Cornell University Press, 1980); Sarah Maza, *Private Lives and Public Affairs: The Causes Célèbres of Prerevolutionary France* (Berkeley: University of California Press, 1993), 265–66.

17. Abbé de Mably, *Entretiens de Phocion*, in *Oeuvres Complètes* (Paris: Bossange, Masson, et Besson, 1797), 10:48.

18. Albert-Joseph Hennet, *Du Divorce* (Paris: Desenne, 1789), v; see also Desan, *Family on Trial*, 67–74.

19. For a more detailed definition of sensibility, see Carolyn Purnell, "Instruments Endowed with Sensibility: Remaking Society through the Body in Eighteenth-Century France" (PhD diss., University of Chicago, 2013).

20. William Reddy, *The Navigation of Feeling: A Framework of the History of Emotions* (Cambridge: Cambridge University Press, 2001), 141–72.

21. David Denby, *Sentimental Narrative and the Social Order in France, 1760–*

*1820* (Cambridge: Cambridge University Press, 1994); Anne C. Vila, *Enlightenment and Pathology: Sensibility in the Literature and Medicine of Eighteenth-Century France* (Baltimore, MD: Johns Hopkins University Press, 1998); Maza, *Myth of the French Bourgeoisie*, 41–68; Maza, *Private Lives and Public Affairs*, 84–85; Lynn Hunt, *Inventing Human Rights: A History* (New York: Norton, 2007), 35–69; Allan H. Pasco, *Revolutionary Love in Eighteenth- and Early Nineteenth-Century France* (Surrey: Ashgate, 2009), 33–96; G. J. Barker-Benfield, *The Culture of Sensibility: Sex and Society in Eighteenth-Century Britain* (Chicago: University of Chicago Press, 1992), 248.

22. Vila, *Enlightenment and Pathology*, 112–50; Maza, *Myth of the French Bourgeoisie*, 41–68; Philip Stewart, *L'Invention du Sentiment: Roman et économie affective au XVIIIe siècle* (Oxford: Studies on Voltaire and the Eighteenth Century, 2010), 175–90.

23. There is a useful historiographical summary of work on companionate marriage in Jeffrey Watt, *The Making of Modern Marriage: Matrimonial Control and the Rise of Sentiment in Neuchâtel, 1550–1800* (Ithaca, NY: Cornell University Press, 1992), 1–23; for more recent work, see also Desan, *Family on Trial*, 67–74; Maurice Daumas, *Le mariage amoureux: Histoire du lien conjugal sous l'Ancien Régime* (Paris: Armand Collin, 2004), 259–91; Dena Goodman, *Becoming a Woman in the Age of Letters* (Ithaca, NY: Cornell University Press, 2009), 322; Kate Retford, *The Art of Domestic Life: Family Portraiture in Eighteenth-Century England* (New Haven, CT: Yale University Press, 2006); Amanda Vickery, *The Gentleman's Daughter: Women's Lives in Georgian England* (New Haven, CT: Yale University Press, 1998); Siân Reynolds, *Marriage and Revolution: Monsieur and Madame Roland* (Oxford: Oxford University Press, 2012).

24. Retford, *Art of Domestic Life*, 8.

25. Mita Choudhury, *Convents and Nuns in Eighteenth-Century French Politics and Culture* (Ithaca, NY: Cornell University Press, 2004), 136; Jacqueline Hecht, "From 'Be Fruitful and Multiply' to Family Planning: The Enlightenment Transition," *Eighteenth-Century Studies* 32, no. 4 (1999): 545; Maza, *Myth of the French Bourgeoisie*, 41–68; Carol Blum, *Strength in Numbers: Population, Reproduction, and Power in Eighteenth-Century France* (Baltimore, MD: Johns Hopkins University Press, 2002), 21–60.

26. Anne C. Vila, "Sex, Procreation, and the Scholarly Life from Tissot to Balzac," *Eighteenth-Century Studies* 35, no. 2 (2002): 240; Michael Winston, "Medicine, Marriage, and Human Degeneration in the French Enlightenment," *Eighteenth-Century Studies* 38, no. 2 (2005): 263–81.

27. Claire Cage, *Unnatural Frenchmen: The Politics of Priestly Celibacy and Marriage, 1720–1815* (Charlottesville, VA: University of Virginia Press, 2015); Claire Cage, "'Celibacy Is a Social Crime': The Politics of Clerical Marriage, 1794–1799," *French Historical Studies* 36, no. 4 (2013): 601–28.

28. Elena Russo, *Styles of Enlightenment: Taste, Politics, and Authorship in Eighteenth-Century France* (Baltimore, MD: Johns Hopkins University Press, 2007), 42; Roger Chartier, "The Man of Letters," in *Enlightenment Portraits*, ed. Michel Vovelle, trans. Lydia Cochrane (Chicago: University of Chicago Press, 1997), 142–89. Accordingly, philosophes became vulnerable to accusations of hypocrisy. Eric Walter, "Le Complexe d'Abelard ou le célibat des gens de lettres," *Dix-Huitième Siècle* 12 (1980): 127–52.

29. Russo, *Styles of Enlightenment*, 1; see also Gregory S. Brown, *A Field of Honor: Writers, Culture, and Public Theater in French Literary Life from Racine to the Revolution* (New York: Columbia University Press/EPIC, 2002), chap. 3.

30. Du Marsais, "Philosopher."

31. François-Marie Arouet de Voltaire, "Men of Letters," in *The Encyclopedia of Diderot & d'Alembert Collaborative Translation Project*, trans. Dena Goodman (Ann Arbor: Michigan Publishing, University of Michigan Library, 2003), http://hdl.handle.net/2027/spo.did2222.0000.052 (accessed 19 June 2014). Originally published as "Gens de lettres," in *Encyclopédie ou Dictionnaire raisonné des sciences, des arts et des métiers* (Paris, 1757), 7:599–600.

32. David A. Bell, *Cult of the Nation: Inventing Nationalism, 1680–1800* (Cambridge, MA: Harvard University Press, 2003), 124; Charles Paul, *Science and Immortality: The Éloges of the Paris Academy of Sciences* (Berkeley: University of California Press, 1980), 14–19.

33. Bell, *Cult of the Nation*, 121–28; Ribard, *Raconter, Vivre, Penser*, 51–56; Jean-Claude Bonnet, *Naissance du Panthéon: Essai sur le Culte des Grands Hommes* (Paris: Fayard, 1998), 29–112; Marisa Linton, *The Politics of Virtue in Enlightenment France* (New York: Palgrave, 2001), 1–9, 51–79.

34. Bernard de Fontenelle, *Eloges des Academiciens avec l'histoire de l'Academie Royale des Sciences en 1699*, 2 vols. (Hay: Isaac Van Der Kloot, 1740), 2:40.

35. Ibid., 2:6.

36. Charles Perrault, *Les Hommes Illustres qui ont paru en France pendant ce siècle avec leurs portraits au naturel* (1697; repr., Geneva: Slatkine Reprint, 1970), 2:57.

37. Fontenelle, *Éloges*, 2:344 (Newton), 2:246–47 (Alexis Littre).

38. Perrault, *Hommes Illustres*, 2:57–58 (Le Fevre), 1:49 (Sainte-Marthe). See also ibid., 1:130 and 1:136–37; Fontenelle, *Éloges*, 2:81 and 2:415.

39. Anne Goldgar, *Impolite Learning: Conduct and Community in the Republic of Letters, 1680–1750* (New Haven, CT: Yale University Press, 1995); Shelford, *Transforming the Republic of Letters*; Anthony Grafton, *Worlds Made by Words: Scholarship and Community in the Modern West* (Cambridge, MA: Harvard University Press, 2009).

40. Goldgar, *Impolite Learning*, 159.

41. Ibid., 151.

42. Sturdy, *Science and Social Status*, 267.

43. René Descartes, *Discourse on the Method and Meditations on First Philosophy*, ed. David Weissman, trans. Elizabeth S. Haldane and G. R. T. Ross (New Haven, CT: Yale University Press, 1996), 3–48; Michel de Montaigne, *The Essays: A Selection*, trans. and ed. M. A. Screech (New York: Penguin Books, 1991), 102; Matthew L. Jones, *The Good Life in the Scientific Revolution: Descartes, Pascal, Leibniz, and the Cultivation of Virtue* (Chicago: Chicago University Press, 2006); Steven Shapin, "'The Mind Is Its Own Place': Science and Solitude in Seventeenth-Century England," *Science in Context* 4, no. 1 (1990): 203–5.

44. Jean le Rond d'Alembert, "Essai sur la société des gens de lettres et des grands," in *Mélanges de Littérature, d'Histoire, et de Philosophie* (Amsterdam: Zacharie Chatelain et Fils, 1764), 1:383–98. On independence and civility, see Steven Shapin, *A Social History of Truth: Civility and Science in Seventeenth-Century England* (Chicago:

University of Chicago Press, 1994); J. B. Shank, *The Newton Wars and the Beginning of the French Enlightenment* (Chicago: University of Chicago Press, 2008); Goldgar, *Impolite Learning*; Russo, *Styles of Enlightenment*.

45. Rémy Saisselin, *The Literary Enterprise in Eighteenth-Century France* (Detroit, MI: Wayne State University Press 1979), 127.

46. Néricault Destouches, *Le Philosophe Marié, ou le marie honteux de l'être* (Paris: Librairie de la Bibliothèque Nationale, 1885), 17, 88.

47. Vila, "Faux Savants, Femmes Philosophes & Philosophes Amoureux," 211–14.

48. M. Doigni Du Ponceau, "Epitre a un Homme de Lettres Célibataire" (Paris: J. B. Brunet, 1773), 1, 4.

49. Paul, *Science and Immortality*, 27.

50. J. A. N. Condorcet, *Oeuvres de Condorcet*, ed. Arthur Condorcet-O'Connor (Paris: Firmin Didot Frères, 1847–1849), 2:659–60, 2:236–37.

51. Ibid., 2:429, 3:271.

52. Cailly, *Griefes et Plaintes des Femmes Mal Mariée* (France, 1789), 6.

53. Condorcet, *Oeuvres*, 3:367.

54. Ibid., 2:430.

55. Louis-Pierre Manuel, *L'année françoise, ou vies des hommes qui ont honoré la France, ou par leurs talens, ou par leurs service et sur-tout par leurs vertus*, 4 vols. (Paris, 1789), 4:286 (Tronchin), 4:398 (Racine).

56. For the continued importance of friendship, see ibid., 2:77.

57. Ibid., 4:101.

58. Condorcet, *Oeuvres*, 2:637–38.

59. Jean-Jacques Garnier, *L'Homme de Lettres* (Paris, 1764), 191–92.

60. Jérôme Lalande quoted in Simone Dumont, *Un Astronome des Lumières: Jérôme Lalande* (Paris: Vuibert, 2007), 149–50.

61. Lalande, "Testament Moral," 52.

62. Jérôme Lalande, *Éloge de M. Commerson* (S.I., n.d.), 5.

63. Ibid., 6.

64. Voltaire quoted in Roger Pearson, *Voltaire Almighty: A Life in Pursuit of Freedom* (New York: Bloomsbury, 2005), 128; see also Judith P. Zinsser, *La Dame d'Esprit: A Biography of the Marquise Du Châtelet* (New York: Viking, 2006); David Bodanis, *Passionate Minds: The Great Enlightenment Love Affair* (New York: Crown, 2006).

65. Judith P. Zinsser, "Entrepreneur of the 'Republic of Letters': Emilie de Breteuil, Marquise du Châtelet, and Bernard Mandeville's Fable of the Bees," *French Historical Studies* 25, no. 4 (Fall 2002): 595–624, 604.

66. Claude-Adrien Helvétius to Comtesse de Rochefort, 19 July 1751, in *Correspondance Générale d'Helvétius*, ed. Peter Allen, Alan Dainard, Marie-Thérèse Inguenaud, Jean Orsoni, and David Smith (Oxford: Voltaire Foundation, 1984), 1:275.

67. Claude-Adrien Helvétius to Ottaviano di Guasco, 31 August 1751, in *Correspondance Générale*, 1:286.

68. Claude-Adrien Helvétius to Anne-Catherine Ligneville Helvétius, n.d. [early 1750s], in *Correspondance Générale*, 1:293.

69. Claude-Adrien Helvétius to Anne-Catherine Ligneville Helvétius, 8 July 1755, in *Correspondance Générale*, 1:329.

70. Claude-Adrien Helvétius to Anne-Catherine Ligneville Helvétius, 26 March 1764, in *Correspondance Générale*, 3:114–15.

71. Claude-Adrien Helvétius to Anne-Catherine Ligneville Helvétius, 19 August 1758, in *Correspondance Générale*, 2:90.

72. Manuel noted approvingly that he was a "good husband, good father, happy with his wife and children, who tasted all the pleasures of domestic life." Manuel, *L'année françoise*, 4:413. Saint-Lambert likewise described him as "a true philosophe, a lovable man, loved and happy"; quoted in Ribard, *Raconter, Vivre, Penser*, 234.

73. Claude-Adrien Helvétius to Charles Palissot de Montenoy, September 1751, in *Correspondance Générale*, 1:288.

74. Nina Kushner, *Erotic Exchanges: The World of Elite Prostitution in Eighteenth-Century Paris* (Cornell: Cornell University Press, 2014), 110; Moussa Baccus, "Helvétius et la sexualité," in *Sexualité, mariage et famille au XVIIIe siècle*, ed. Olga Cragg and Rosena Davison (Quebec: Presse de la Université Laval, 1998), 56. On aristocratic libertinage, see Lenard R. Berlanstein, *Daughters of Eve: A Cultural History of French Theater Women from the Old Regime to the Fin de Siècle* (Cambridge, MA: Harvard University Press, 2001), chap. 2.

75. Melchior Grimm to General Betzki, 10 February 1765, in Diderot, *Correspondance*, ed. George Roth (Paris, 1955–1970), 5:25. The letter was most likely written in consultation, as the quotation came from one of Diderot's letters recounting the contents of Grimm's letter.

76. Edward G. Andrew, *Patrons of Enlightenment* (Toronto: University of Toronto Press, 2006), 10–12.

77. For similar representations of men of letters and marriage, see Retford, *Art of Domestic Life*, 141–47.

78. General Betzki to Melchior Grimm, 16 March 1765, in Diderot, *Correspondance*, 5:26.

79. Correspondence between J. B. A. Suard and Amélie Panckouke, c. 1764–1765, included in L. A. Boiteux, "Voltaire et le ménage Suard," in *Travaux sur Voltaire et le Dix-Huitième Siècle*, ed. Theodore Besterman (Geneva: Institut et Musee Voltaire, 1955), 34.

80. Pierre Choderlos de Laclos to Marie-Soulange Laclos, 2 messidor an 2, in *Lettres inédites*, 72.

81. Pierre Choderlos de Laclos to Marie-Soulange Laclos, 29 prairial an 2, in *Lettres inédites*, 69–70.

82. Pierre Choderlos de Laclos to Marie-Soulange Laclos, 14 prairial an 2, in *Lettres inédites*, 60.

83. Pierre Choderlos de Laclos to Marie-Soulange Laclos, 12 fructidor an 2, in *Lettres inédites*, 86.

84. Pierre Choderlos de Laclos to Marie-Soulange Laclos, 26 thermidor an 8, in *Lettres inédites*, 142.

85. "Journal d'Ampère et prière à la mort de Julie," Papiers de A. M. Ampère, Cart 23, Chap. 20, Chemise 327, Archives de l'Académie de Sciences, Paris, France.

86. Saint-Preux, in *La Nouvelle Héloïse*, quoted in Denby, *Sentimental Narrative*, 103.

87. Goodman, *Becoming a Woman*, 322; Denby, *Sentimental Narrative*, 115.

88. Condorcet to Amélie Suard, May 1771, in *Correspondance Inédite de Condorcet et Mme. Suard, M. Suard, et Garat (1771–1791)*, ed. Elisabeth Badinter (Paris: Fayard, 1988), 48.

89. Even as Helvétius acknowledged his infidelities, he promised his wife that he loved her best. He apparently saw little conflict between the personae of loving husband and gallant libertine. Baccus, "Helvétius et la sexualité," 56.

90. This idea is best expressed in Jean-Jacques Rousseau, *Emile, ou de l'éducation*, ed. Michel Launay (Paris: GF Flammarion, 1966), 35–52, 524–25; A. L. Thomas, Denis Diderot, and Louise d'Epinay, in *Qu'est-ce qu'une femme?*, ed. Elisabeth Badinter (Paris: P.O.L., 1989). See also Goodman, *Becoming a Woman*, 301–2; Desan, *Family on Trial*, 35–40.

91. Female members of the household writ large were also put to work. See Mary Terrall, "Frogs on the Mantelpiece: The Practice of Observation in Daily Life," in *Histories of Scientific Observation*, ed. Lorraine Daston and Elizabeth Lunbeck (Chicago: University of Chicago Press, 2011); Patricia Fara, *Pandora's Breeches: Women, Science, and Power in the Enlightenment* (London: Pimlico, 2004); Dorinda Outram and Pnina G. Abir-Am, eds., *Uneasy Careers and Intimate Lives: Women in Science, 1789–1979* (Newark, NJ: Rutgers University Press, 1987); Reynolds, *Marriage and Revolution*, 79–87.

92. Jean de Magalhaens to Antoine Lavoisier, 13 November 1775, in *Oeuvres de Lavoisier: Correspondance*, ed. René Fric (Paris: Michel Albion, 1957), 2:505; Arthur Young quoted in Fara, *Pandora's Breeches*, 176.

93. Antoine Lavoisier quoted in Jean-Pierre Poirier, *Lavoisier: Chemist, Biologist, Economist*, trans. Rebecca Balinski (Philadelphia: University of Pennsylvania Press, 1996), 358.

94. Marco Beretta, *Imaging a Career in Science: The Iconography of Antoine Laurent Lavoisier* (Canton, MA, 2001), 41–42.

95. "[Women] . . . provided the social setting in which future protégés could collect within the ambit of their patron." Dorinda Outram, "Before Objectivity: Wives, Patronage, and Cultural Reproduction in Early Nineteenth-Century French Science," in Outram and Abir-Am, *Uneasy Careers and Intimate Lives*, 29. See also Beretta, *Imaging a Career*.

96. Madame Diderot declined to attend a production of *Père de Famille* until it had nearly closed; Diderot wrote to Sophie Volland in September 1769, "Enfin madame Diderot prit, le vendredi au soir, la veille de la dernière représentation, le parti d'y aller avec sa fille. Elle sentit l'indécence qu'il y avoit à répondre à tous ceux qui lui faisoient compliment, qu'elle n'y avoit pas été." *Correspondance*, 9:136.

97. Darnton, "A Police Inspector Sorts His Files," 169–72.

CHAPTER TWO

1. Mary Terrall, *Catching Nature in the Act: Réaumur and the Practice of Natural History in the Eighteenth Century* (Chicago: University of Chicago Press, 2014), 20–78; Londa Schiebinger, *The Mind Has No Sex? Women in the Origins of Modern Science*

(Cambridge, MA: Harvard University Press, 1991), 66–101; Deborah Harkness, "Managing an Experimental Household: The Dees of Mortlake and the Practice of Natural Philosophy," *Isis* 88, no. 2 (1997): 247–62; Alix Cooper, "Homes and Households," in *The Cambridge History of Science*, vol. 3, *Early Modern Science*, ed. Katharine Park and Lorraine Daston (Cambridge: Cambridge University Press, 2006), 224–37; Monika Mommertz, "The Invisible Economy of Science: A New Approach to the History of Gender and Astronomy at the Eighteenth-Century Berlin Academy of Sciences," trans. Julia Barker, in *Men, Women, and the Birthing of Modern Science*, ed. Judith P. Zinsser (DeKalb: Northern Illinois University Press, 2005), 159–78; Helena M. Pycior, Nancy G. Slack, and Pnina Abir-Am, eds., *Creative Couples in the Sciences* (New Brunswick, NJ: Rutgers University Press, 1996).

2. Mathematicians in particular considered imagination and excess sensibility (both gendered feminine) to be intellectual liabilities. Mary Terrall, "Metaphysics, Mathematics, and the Gendering of Science in Eighteenth-Century France," in *The Sciences in Enlightened Europe*, ed. William Clark, Jan Golinski, and Simon Schaffer (Chicago: University of Chicago Press, 1999), 246–71, esp. 257–60.

3. Amanda Vickery, *Behind Closed Doors: At Home in Georgian England* (New Haven, CT: Yale University Press, 2009), 49–82.

4. Richard Holmes, *Age of Wonder: The Romantic Generation and the Discovery of the Beauty and Terror of Science* (New York: Vintage, 2010), 81–124; Gadi Algazi, "Scholars in Households: Refiguring the Learned Habitus, 1480–1550," *Science in Context* 16, no. 1 (2003): 11–18; Patricia Fara, *Pandora's Breeches: Women, Science, and Power in the Enlightenment* (London: Pimlico, 2004), 107–86.

5. Robert Shackleton, *Montesquieu: A Critical Biography* (Oxford: Oxford University Press, 1961), 79.

6. Accounting was one of the most essential skills a woman of Baronne de Montesquieu's station needed to master. Vickery, *The Gentleman's Daughter*, 127–60; Shackleton, *Montesquieu*, 14–15, 79, 194, and esp. 197–98; Madame de Montesquieu to Sarrau de Vésis, 25 May 1723, and Madame de Montesquieu to Lalanne, 11 April 1726, in Charles de Secondat, baron de Montesquieu, *Correspondance de Montesquieu*, ed. François Gebelin and André Morize (Paris: Édouard Champion, 1914), 1:23 and 132–33.

7. Madame de Montesquieu to Grégoire, 11 March 1737, in *Correspondance de Montesquieu*, 1:328–29.

8. Letters from Julie Carron to A. M. Ampère, 26 February 1802, 23 April 1802, in "Lettres de Julie Carron première femme," Papiers de A. M. Ampère, Cart 23, Chap. 20, Chemise 331, Archives de l'Académie des Sciences, Paris, France.

9. James R. Hofmann, *André-Marie Ampère* (Oxford: Blackwell, 1995), 32–34.

10. For theoretical treatments of family labor in scholarly settings, see Algazi, "Scholars in Households"; and Bonnie Smith, *Gender of History: Men, Women, and Historical Practice* (Cambridge, MA: Harvard University Press, 1998), 84–102.

11. Simone Dumont, *Un Astronome des Lumières: Jérôme Lalande* (Paris: Vuibert, 2007), 21, 67, 91, 148; Fara, *Pandora's Breeches*, 18.

12. Elisabeth Badinter, "Un couple d'astronomes: Jérôme Lalande et Reine Lepaute," in *Société Archéologique, Scientifique, et Littéraire de Béziers*, 10th ed., vol. 1 (Beziers: Hérault, 2004–2005), 71–72; Dumont, *Un Astronome des Lumières*, 22–23.

13. Badinter, "Un couple d'astronomes," 72–74; Dumont, *Un Astronome des Lumières*, 91.

14. Charlotte de Saxe-Gotha to Amélie Lefrancais Lalande, 23 December 1798, and Charlotte de Saxe-Gotha to Jérôme Lalande, 8 Floreal l'An 7, Lettres de la duchesse Charlotte de Saxe-Gotha, copies de Julien Raspail, MS 2761, Bibliothèque Inguimbertine, Carpentras, France.

15. See, for example, Carol Pal, *Republic of Women: Rethinking the Republic of Letters in the Seventeenth Century* (Cambridge: Cambridge University Press, 2012), 67.

16. Dumont, *Un Astronome des Lumières*, 327; Jérôme Lalande to Louise-Elisabeth-Félicité Du Pierry, 15 messidor an 2, in Jérôme Lalande, *Lettres à Madame du Pierry et au Juge Honoré Flaugergues* (Paris: Vrin, 2007), 54.

17. Jérôme de Lalande, *Bibliographie Astronomique avec l'Histoire de l'Astronomie, depuis 1781 jusqu'à 1802* (Paris: l'Imprimerie de la République, 1803), 691.

18. Joseph Jérôme Lefrançois de Lalande, "Dotation faite à son neveu et à sa nièce," Documents pour une biographie de Lalande: pièces écrites par l'astronome lui-même, MS 2763, Bibliothèque Inguimbertine, Carpentras, France.

19. On developments that transformed calculating into something lesser, see Lorraine Daston, "Enlightenment Calculations," *Critical Inquiry* 21, no. 1 (1994): 182–202; on calculation as gendered masculine during the eighteenth century, see Terrall, "Metaphysics, Mathematics, and the Gendering of Science," 257–60; Alison Winter, "The Calculus of Suffering: Ada Lovelace and the Corporeal Constraints on Women's Knowledge in Early Victorian England," in *Science Incarnate: The Physical Presentation of Intellectual Selves*, ed. Christopher Lawrence and Steven Shapin (Chicago: University of Chicago Press, 1997), 207. On a lack of intelligence as a desirable quality in calculators, see Simon Schaffer, "Babbage's Intelligence: Calculating Engines and the Factory System," *Critical Inquiry* 21, no. 1 (August 1994): 203–27.

20. Joseph Jérôme Lefrançois de Lalande, "Dotation faite à son neveu et à sa nièce," Documents pour une biographie de Lalande: pièces écrites par l'astronome lui-même, MS 2763, Bibliothèque Inguimbertine, Carpentras, France.

21. Joseph Jérôme Lefrançois de Lalande, "Declaration des biens, an II," MS 2761, Bibliothèque Inguimbertine, Carpentras, France.

22. Jérôme Lalande to Louise-Elisabeth-Félicité Du Pierry, 1 messidor an 2, in *Lettres*, 54.

23. Schiebinger, *Mind Has No Sex?*, 79–101; Claire Brock, *The Comet Sweeper: Caroline Herschel's Astronomical Ambition* (Cambridge: Icon Books, 2007), chaps. 3 and 4; Fara, *Pandora's Breeches*, chaps. 7 and 8; Rob Iliffe and Frances Willmoth, "Astronomy and the Domestic Sphere: Margaret Flamsteed and Caroline Herschel as Assistant Astronomers," in *Women, Science, and Medicine, 1500–1700: Mothers and Sisters of the Royal Society*, ed. Lynette Hunter and Sarah Hutton (Thrupp, UK: Sutton Publishing, 1997), 235–65.

24. Jean de Magalhaens to Antoine Lavoisier, 13 November 1775, in *Oeuvres de Lavoisier: Correspondance*, ed. René Fric (Paris: Michel Albion, 1957), 2:505.

25. Jean-Pierre Poirier, *Lavoisier: Chemist, Biologist, Economist*, trans. Rebecca Balinski (Philadelphia: University of Pennsylvania Press, 1996), 39–40.

26. Hassenfratz to Madame Lavoisier, 18 March 1787, Landriani to Madame

Lavoisier, 12 October 1788, Born to Madame Lavoisier, 29 January 1791, and Landriani to Madame Lavoisier, 14 September 1791, in *Oeuvres de Lavoisier*, 5:25, 5:219, 6:228, and 6:323.

27. Poirier, *Lavoisier*, 95–96.

28. Bruno Belhoste, *Paris Savant: Parcours et rencontres au temps des Lumières* (Paris: Armand Colin, 2011), 213.

29. Marie-Anne Paulze Lavoisier to her brother, 30 August 1777, in *Oeuvres de Lavoisier*, 3:605.

30. Anders-Johann Lexell quoted in Poirier, *Lavoisier*, 96.

31. Poirier, *Lavoisier*, 245.

32. Marco Beretta, *Imaging a Career in Science: The Iconography of Antoine Laurent Lavoisier* (Canton, MA: Watson Publishing International, 2001), 44.

33. Translating and advocating were typical activities for women involved in science. See Georges Cuvier, "Materiaux pour l'éloge de Guyton de Morveau," Éloges des savants, Papiers et correspondance du baron George Cuvier, MS 3144, Bibliothèque de l'Institut, Paris, France; Paula Findlen, "Translating the New Science: Women and the Circulation of Knowledge in Enlightenment Italy," *Configurations* 3 (1995): 167–206; Dorinda Outram, "Before Objectivity: Wives, Patronage, and Cultural Reproduction in Early Nineteenth-Century French Science," in *Uneasy Careers and Intimate Lives: Women in Science, 1789–1979*, ed. Pnina G. Abir-Am and Dorinda Outram (Newark, NJ: Rutgers University Press, 1987).

34. Josiah Wedgwood to Antoine Lavoisier, 18 August 1791, trans. Marie Lavoisier, and Antoine Lavoisier to Josiah Wedgwood, 12 September 1791, in *Oeuvres de Lavoisier*, 6:314–16 and 323.

35. Hassenfratz to Madame Lavoisier, 20 February 1788, in *Oeuvres de Lavoisier*, 5:135.

36. Robert Siegfried, "Lavoisier and the Phlogistic Connection," *Ambix* 36 (1989): 31–40; for more on Lavoisier's innovations, especially with regard to language, see Jessica Riskin, *Science in the Age of Sensibility: The Sentimental Empiricists of the French Enlightenment* (Chicago: University of Chicago Press, 2002), 248–49.

37. Belhoste, *Paris Savant*, 208.

38. For more on Priestley's persona, see Jan Golinski, *Science as Public Culture: Chemistry and Enlightenment in Britain, 1760–1820* (Cambridge: Cambridge University Press, 1992), 86–90.

39. Horace Benedict de Saussure to Madame Lavoisier, 28/29 February 1788, in *Oeuvres de Lavoisier*, 5:139–40.

40. Ibid., 5:140.

41. Marsilio Landriani to La Rochefoucauld, 23 September 1788, and Marsilio Landriani to Madame Lavoisier, 12 October 1788, in *Oeuvres de Lavoisier*, 5:216 and 219.

42. Horace Benedict de Saussure to Madame Lavoisier, 7 November 1788, in *Oeuvres de Lavoisier*, 5:232.

43. Ibid.

44. Jao Jacinthe de Magellan to Madame Lavoisier, May 1775, and 16 January 1776, in *Oeuvres de Lavoisier*, 2:508 and 3:537.

45. Marie-Anne Lavoisier, "Notes sur Lavoisier," Éloges des savants, Papiers et correspondance du baron George Cuvier, MS 3145, Bibliothèque de l'Institut, Paris, France.

46. She also painted a portrait of Benjamin Franklin in 1788. Madeleine Pinault Sörensen, "Madame Lavoisier, dessinatrice et peintre," *La Revue du Musée des Arts et Métiers* 6 (1994): 23–25.

47. Mary Vidal, "The 'Other Atelier': Jacques-Louis David's Female Students," in *Women, Art, and the Politics of Identity in Eighteenth-Century Europe*, ed. Melissa Hyde and Jennifer Milam (Aldershot, UK: Ashgate Publishing, 2003), 243–44.

48. Arthur Donovan, *Antoine Lavoisier: Science, Administration, and Revolution* (Cambridge: Cambridge University Press, 1993), 278.

49. Count Rumford quoted in Poirier, *Lavoisier*, 400.

50. Poirier, *Lavoisier*, 401.

51. Dorinda Outram, "The Language of Natural Power: The 'Eloges' of Georges Cuvier and the Public Language of Nineteenth-Century Science," *History of Science* 16 (1978): 153–78.

52. Léonard Burnand, *Necker et l'opinion publique* (Paris: Honoré Champion Éditeur, 2004); Jean-Denis Bredin, *Une singulière famille: Jacques Necker, Suzanne Necker, et Germaine de Staël* (Paris: Fayard, 1999), 43–75.

53. Cissie C. Fairchilds, *Poverty and Charity in Aix-en-Provence, 1640–1789* (Baltimore, MD: Johns Hopkins University Press, 1976), 81.

54. Laurence Brockliss and Colin Jones, *The Medical World of Early Modern France* (Oxford: Clarendon Press, 1997), 726.

55. Stuart Woolf, "The *Société de Charité Maternelle*, 1788–1815," in *Medicine and Charity before the Welfare State*, ed. Jonathan Barry and Colin Jones (London: Routledge, 1991), 102–5.

56. Brockliss and Jones, *Medical World*, 727.

57. Jacques Necker, *Compte Rendu par M. Necker, Directeur général des Finances, au mois de janvier 1781, imprimé par ordre de Sa Majesté* (repr., London: G. Kearsley, 1781), 127.

58. *Mémoires secrets pour servir à l'histoire de la république des lettres en France* (London: chez John Adamson, 1781–1789), 14:169.

59. *Mémoires secrets*, 14:186.

60. Léonard Burnard, *Les Pamphlets Contre Necker: Médias et imaginaire politique au XVIIIe siècle* (Paris: Éditions Classiques Garnier, 2009), 134–42.

61. Sonja Boon, "Performing the Woman of Sensibility: Suzanne Curchod Necker and the Hospice de Charité," *Journal for Eighteenth-Century Studies* 35, no. 2 (June 2009): 235–54.

62. Burnard, *Les Pamphlets Contre Necker*, 20–22; Dena Goodman, *The Republic of Letters: A Cultural History of the French Enlightenment* (Ithaca, NY: Cornell University Press, 1994).

63. Suzanne Necker, "Portrait de M. Necker fait en 1787," in *Mélanges extraits des manuscrits de Mme Necker* (Paris: C. Pougens, 1798), 2:373, 375, 380, 381–86, 378.

64. Léonard Burnand, *Necker et l'opinion publique* (Paris: Honoré Champion Éditeur, 2004), 32–33.

65. Germaine de Staël quoted in Dena Goodman, "Suzanne Necker's *Mélanges*: Gender, Writing, and Publicity," in *Going Public: Women in Publishing in Early*

*Modern France*, ed. Elizabeth C. Goldsmith and Dena Goodman (Ithaca, NY: Cornell University Press, 1995), 210–20, 210.

66. Ian Cumming, *Helvétius: His Life and Place in the History of Educational Thought* (London: Routledge, 2001), 82–84; for more on antiphilosophe discourses, see Darrin McMahon, *Enemies of the Enlightenment: The French Counter-Enlightenment and the Making of Modernity* (New York: Oxford University Press, 2001), 18–53.

67. Claude-Adrien Helvétius to Anne-Catherine Ligneville Helvétius, 7 July 1758, in *Correspondance Générale d'Helvétius*, ed. Peter Allen, Alan Dainard, Marie-Thérèse Inguenaud, Jean Orsoni, and David Smith (Oxford: Voltaire Foundation, 1984), 2:49.

68. Claude-Adrien Helvétius to Anne-Catherine Ligneville Helvétius, 9 Aug 1758, in *Correspondance Générale*, 2:73.

69. Anne-Catherine Ligneville Helvétius, [speech to Parlement], January–February 1759, in *Correspondance Générale*, 2:430–31.

70. Anne-Catherine Ligneville Helvétius to Malesherbes, 24 August 1759, in *Correspondance Générale*, 2:266–67.

71. Anne-Catherine Ligneville Helvétius to Jean Lévesque de Burigny, 18 May 1760, in *Correspondance Générale*, 2:277–78.

72. Anne-Catherine Ligneville Helvétius to Jean Lévesque de Burigny, 29 June 1760, in *Correspondance Générale*, 2:283–84.

73. J. B. Shank, *The Newton Wars and the Beginning of the French Enlightenment* (Chicago: University of Chicago Press, 2008), 304–5, 321; McMahon, *Enemies of the Enlightenment*, 7–8.

74. W. A. Smeaton, "Madame Lavoisier, P. S. and E. I. Du Pont de Nemours, and the Publication of Lavoisier's 'Mémoires de chimie,'" *Ambix* 36 (1989): 2–30; Madame Condorcet-O'Connor, "Observations rélatives à Condorcet Histoire des Girondins," Nouvelles Acquisitions Françaises, MS 1374, Bibliothèque Nationale de France, site Richelieu, Paris, France; Angélique Vandeul, "Mémoires pour servir à l'histoire de sa vie et de ses ouvrages de M. Diderot par Madm. A. Vandeul, sa fille," Nouvelles Acquisitions Françaises MS 9216, Bibliothèque Nationale de France, site Richelieu, Paris, France.

75. G. J. Barker-Benfield, *The Culture of Sensibility: Sex and Society in Eighteenth-Century Britain* (Chicago: University of Chicago Press, 1992), 248; David Denby, *Sentimental Narrative and the Social Order in France, 1760–1820* (Cambridge: Cambridge University Press, 1994); Vila, *Enlightenment and Pathology*; Sarah Maza, *The Myth of the French Bourgeoisie: An Essay on the Social Imaginary, 1750–1850* (Cambridge, MA: Harvard University Press, 2003), 41–68; Sarah Maza, *Private Lives and Public Affairs: The Causes Célèbres of Pre-revolutionary France* (Berkeley: University of California Press, 1993), 84–85.

76. Terrall, *Catching Nature in the Act*, 30.

77. Pal, *Republic of Women*; Paula Findlen, "Becoming a Scientist: Gender and Knowledge in Eighteenth-Century Italy," *Science in Context* 16, no. 1 (March 2003): 59–87; Massimo Mazzotti, *The World of Maria Gaetana Agnesi, Mathematician of God* (Baltimore, MD: Johns Hopkins University Press, 2007); Natalie Zemon Davis, *Women on the Margins: Three Seventeenth-Century Lives* (Cambridge, MA: Harvard University Press, 1995), 140–202; Ann T. Shteir, *Cultivating Women, Cultivating Science:*

*Flora's Daughters and Botany in England, 1760–1820* (Baltimore, MD: Johns Hopkins University Press, 1996), 33–77.

78. On Du Châtelet, see Judith P. Zinsser, *La Dame d'Esprit: A Biography of the Marquise Du Châtelet* (New York: Viking, 2006); Judith P. Zinsser and Julie Chandler Hayes, eds., *Émilie Du Châtelet: Rewriting Enlightenment Philosophy and Science* (Oxford: Voltaire Foundation, 2006); Robyn Arianrhod, *Seduced by Logic: Émilie Du Châtelet, Mary Sommerville, and the Newtonian Revolution* (New York: Oxford University Press, 2012), 67–146; Mary Terrall, "Émilie Du Châtelet and the Gendering of Science," *History of Science* 33, no. 3 (1995): 283–310. On Bassi, see Paula Findlen, "Science as a Career in Enlightenment Italy: The Strategies of Laura Bassi," *Isis* 84 (1993): 441–69; Gabriella Berti Logan, "The Desire to Contribute: An Eighteenth-Century Italian Woman of Science," *American Historical Review* 99 (1994): 785–812.

79. Terrall, "Émilie Du Châtelet and the Gendering of Science," 289–94; Mary Terrall, "Gendered Spaces, Gendered Audience: Inside and Outside the Paris Academy of Science," *Configurations* 3, no. 2 (1995): 207–32, esp. 223–30.

80. Paula Findlen, "The Scientist's Body," in *The Faces of Nature in Enlightenment Europe*, ed. Lorraine Daston and Gianna Pomata (Berlin: Berliner-Wissenschafts-Verlag, 2003), 211–36.

81. Paola Bertucci, "The In/visible Woman: Mariangela Ardinghelli and the Circulation of Knowledge between Paris and Naples in the Eighteenth Century," *Isis* 104, no. 2 (June 2013): 226–49, esp. 244–49.

82. Jean-Jacques Rousseau, *Emile, ou de l'éducation*, ed. Michel Launay (Paris: GF Flammarion, 1966), bk. 5.

83. For example: "La nature a mis d'un côté la force & la majesté, le courage & la raison; de l'autre, les graces & la beauté, la finesse & le sentiment. Ces avantages ne sont pas toûjours incompatibles; ce sont quelquefois des attributs différens qui se servent de contré poids, ce sont quelquefois les mêmes qualités, mais dans un degré différent. Ce qui est agrément ou vertu dans un sexe, est defaut ou difformité dans l'autre. Les différences de la nature devoient en mettre dans l'éducation; c'est la main du statuaire qui pouvoit donner tant de prix à un morceau d'argile." Desmahis, "Femme (morale)," in *Encyclopédie, ou dictionnaire raisonné des sciences, des arts et des métiers, etc*, ed. Denis Diderot and Jean le Rond d'Alembert, ARTFL Encyclopédie Project, ed. Robert Morrissey, University of Chicago, Spring 2013, http://encyclopedie .uchicago.edu/.

84. Boudier de Villemert, *Ami des Femmes* (London: 1775), 19, 28, 36.

85. On women's education more generally, see Jennifer J. Popiel, *Rousseau's Daughters: Domesticity, Education, and Autonomy in Modern France* (Durham: University of New Hampshire Press, 2008); Martine Sonnet, *L'éducation des filles au temps des Lumières* (Paris: Les Editions du Cerf, 1987); Sylvia Tomaselli, "The Enlightenment Debate on Women," *History Workshop* 20 (1985): 101–24; Jean Bloch, "Discourses of Female Education in the Writings of Eighteenth-Century French Women," in *Women, Gender and Enlightenment*, ed. Barbara Taylor and Sarah Knott (London: Palgrave Macmillan, 2005).

86. There is a voluminous scholarly literature on this shift, but a few salient texts include Schiebinger, *The Mind Has No Sex?*, 189–244; Lieselotte Steinbrügge, *The*

*Moral Sex: Women's Nature in the French Enlightenment*, trans. Pamela E. Selwyn (Oxford: Oxford University Press, 1995), 18–53; Michael E. Winston, *From Perfectibility to Perversion: Meliorism in Eighteenth-Century France* (New York: Peter Lang, 2005), 84–119; Anne C. Vila, *Enlightenment and Pathology: Sensibility in the Literature and Medicine of Eighteenth-Century France* (Baltimore, MD: Johns Hopkins University Press, 1998), 269; Ludmilla Jordanova, "Natural Facts: An Historical Perspective on Science and Sexuality," in *Nature, Culture and Gender*, ed. C. P. MacCormack and M. Strathern (Cambridge: Cambridge University Press, 1980), 42–69.

87. On the idea that learned women were perceived as neglecting domestic duties, see Frances Harris, "Living in the Neighbourhood of Science: Mary Evelyn, Margaret Cavendish and the Greshamites," and Iliffe and Willmoth, "Astronomy and the Domestic Sphere," in Hunter and Hutton, *Women, Science, and Medicine*, 209 and 237–39.

88. For eighteenth-century strategies, see Findlen, "Science as a Career in Enlightenment Italy"; John R. Iverson, "A Female Member of the Republic of Letters: Du Châtelet's Portrait in the *Bilder-Sal [ . . . ] brümhter Schrifftsteller*," in Zinsser and Hays, *Émilie Du Châtelet*, 35–51, esp. 41–45. For earlier iterations, see Sarah Ross, *The Birth of Feminism: Woman as Intellect in Renaissance Italy and England* (Cambridge, MA: Harvard University Press, 2009); and Pal, *Republic of Women*.

89. Popiel, *Rousseau's Daughters*, 91–103; Denise Z. Davidson, "*Bonnes lectures*: Improving Women and Society through Literature in Post-Revolutionary France," in *The French Experience from Republic to Monarchy, 1793–1824: New Dawns in Politics, Knowledge, and Culture*, ed. Máire F. Cross and David Williams (New York: Palgrave, 2000), 155–67; Anne Verjus, "Gender, Sexuality, and Political Culture," in *A Companion to the French Revolution*, ed. Peter McPhee (Chichester, UK: Wiley-Blackwell, 2012), 206.

90. Louis-Pierre Manuel, *L'année françoise, ou vies des hommes qui ont honoré la France, ou par leurs talens, ou par leurs service et sur-tout par leurs vertus*, 4 vols. (Paris, 1789), 4:276.

91. Georges Cuvier, "Materiaux pour l'éloge de Guyton de Morveau," Éloges des savants, Papiers et correspondance du baron George Cuvier, MS 3144, Bibliothèque de l'Institut, Paris, France.

92. J. A. N. Condorcet, *Oeuvres de Condorcet*, ed. Arthur Condorcet-O'Connor (Paris: Firmin Didot Frères, 1847–1849), 2:320.

93. Manuel, *L'année françoise*, 1:150.

94. Jérôme de Lalande, *Astronomie des Dames* (Paris: Chez Bidault, 1806), 5–6.

95. He may well have considered her "useful but secondary," to borrow Judith Zinsser's phrase. Judith P. Zinsser, introduction to Zinsser, *Men, Women, and the Birthing of Modern Science*, 6.

96. Lalande, *Bibliographie Astronomique*, 679.

97. Ibid., 680, 676.

98. Ibid., 680, 679.

99. Ibid., 691.

100. Ibid., 597, 697, 755, 756.

101. Ibid., 739, 851.

102. On female calculators, see Jessica Riskin, "The Defecating Duck, or, the Ambiguous Origins of Artificial Life," *Critical Inquiry* 29, no. 4 (Summer 2003): 629–31;

Lorraine Daston and Peter Galison, *Objectivity* (New York: Zone Books, 2007); Daston, "Enlightenment Calculations," 182–202.

103. Lalande, *Bibliographie Astronomique*, 703.

104. Ibid., 697.

105. Joseph Jérôme Lefrançois de Lalande, "Pour ma fille," MS 2761, Correspondance de Joseph Jérôme Lefrançois de Lalande, Bibliothèque Inguimbertine Carpentras, France.

106. Ross, *The Birth of Feminism*, 7.

107. Mary Terrall, "Frogs on the Mantelpiece: The Practice of Observation in Daily Life," in *Histories of Scientific Observation*, ed. Lorraine Daston and Elizabeth Lunbeck (Chicago: University of Chicago Press, 2011), 185–205; Mary Terrall, "The Uses of Anonymity in the Age of Reason," in *Scientific Authorship: Credit and Intellectual Property in Science*, ed. Mario Biagioli and Peter Galison (London: Routledge, 2003), 91–112; Sara Stidstone Gronim, "What Jane Knew: A Woman Botanist in the Eighteenth Century," *Journal of Women's History* 19, no. 3 (2007): 33–59; Elisabeth Badinter, *Emilie, Emilie: L'ambition féminine au XVIIIe siècle* (Paris: Flammarion, 1993), 7–37; Erica Harth, *Cartesian Women: Versions and Subversions of Rational Discourse in the Old Regime* (Ithaca: Cornell University Press, 1992), 15–62.

108. On selective anonymity, see Bertucci, "The In/visible Woman," 243–47; and Terrall, *Catching Nature in the Act*, 57.

109. Lalande, *Astronomie des Dames*, 7.

110. Marina Benjamin, "Elbow Room: Women Writers on Science, 1790–1840," in *Science and Sensibility: Gender and Scientific Enquiry, 1780–1945*, ed. Marina Benjamin (Oxford: Basil Blackwell, 1991), 27–59; Terrall, *Catching Nature in the Act*, 57.

CHAPTER THREE

1. The literature on smallpox inoculation is robust. Scholars have considered why the French were slow to inoculate in comparison with the English, the development of new quantitative methods to demonstrate the efficacy of inoculation, and the rise of new ideas of the body politic and public health. See Donald R. Hopkins, *The Greatest Killer: Smallpox in History* (Chicago: University of Chicago Press, 2002); Andrea Rusnock, *Vital Accounts: Quantifying Health and Population in Eighteenth-Century England and France* (Cambridge: Cambridge University Press, 2002); Harry Marks, "When the State Counts Lives: Eighteenth-Century Quarrels over Inoculation," in *Body Counts: Medical Quantification in Historical and Sociological Perspective*, ed. Gérard Jorland, George Weisz, and Annick Opinel (Quebec: McGill-Queen's University Press, 2005), 51–64; Elise Lipkowitz, "Matters of Family, Matters of State: A Cultural History of Inoculation in France, 1754–1774" (master's thesis, Northwestern University, 2003).

2. E.g., J. J. Menuret, *Avis aux Meres sur la Petite Vérole et la Rougeole* (Lyon: Chez les Freres Periss, 1770); *Avis aux Meres au Sujet de l'Inoculation ou Lettre a une Dame de Province, qui hésitait de faire inoculer ses Enfans* (Paris: Chez Des Ventes de la Doué, 1775); F. E. L'Haridon, *Avis aux dames françaises sur l'inoculation de leurs enfans* (Paris: Gabon, 1800).

3. Carol Blum, *Strength in Numbers: Population, Reproduction, and Power in Eighteenth-Century France* (Baltimore, MD: Johns Hopkins University Press, 2002), 3.

4. Lady Wortley Montagu to Mrs. S.C. [Miss Sarah Chiswell], 1 April 1717, in *Letters of the Right Honourable Lady Wortley Montagu*, 3rd ed. (London: T. Becket and P. A. De Hondt, 1763), 2:61.

5. Voltaire, *Letters Concerning the English Nation* (London: C. Davis and A. Lyon, 1733), 69.

6. Lawrence Brockliss and Colin Jones, *The Medical World of Early Modern France* (Oxford: Clarendon Press, 1997), 470–71. For more on the shift away from iatromechanism, see Anne C. Vila, *Enlightenment and Pathology: Sensibility in the Literature and Medicine of Eighteenth-Century France* (Baltimore, MD: Johns Hopkins University Press, 1998), 43–45.

7. Louise Dorothea von Meiningen, Duchesse de Saxe-Gotha, to Voltaire, 28 April 1759, accessible via Electronic Enlightenment, http://dx.doi.org/10.13051/ee:doc/voltfrVF1040136a1c (accessed 17 September 2015).

8. Marie Louise Denis to Pierre Robert le Cornier de Cideville, 4 August 1759, accessible via Electronic Enlightenment, http://dx.doi.org/10.13051/ee:doc/voltfrVF1040293a1c (accessed 17 September 2015).

9. Marie Madeleine de Brémond d'Ars, marquise de Verdelin, to Jean-Jacques Rousseau, 26 September 1762, available via Electronic Enlightenment, http://dx.doi.org/10.13051/ee:doc/rousjeVF0130123a1c (accessed 17 September 2015).

10. Genevieve Miller, *The Adoption of Inoculation for Smallpox in England and France* (Philadelphia: University of Pennsylvania Press, 1957), 234.

11. Victoria Nicole Meyer, "Divining the Pox: The Controversy over Smallpox Inoculation in Eighteenth-Century France" (PhD diss., University of Virginia, 2010), 103–97.

12. "Ode a Monseigneur le Duc d'Orléans sur l'Inoculation" (1756), p. 6, Bibliothèque Nationale de France.

13. Kathleen Wellman, *La Mettrie: Medicine, Philosophy, and Enlightenment* (Durham, NC: Duke University Press, 1992), 92–95; Meyer, "Divining the Pox," 188–91.

14. Voltaire, *Letters*, 70.

15. Voltaire to Joseph Michel Antoine Servan, 13 April 1766, available via Electronic Enlightenment, http://dx.doi.org/10.13051/ee:doc/voltfrVF1140176a1c (accessed 17 September 2015).

16. Denis Diderot, "Insertion de la Petite Vérole," Louis Jaucourt, "Visage," Etienne-Noël Damilaville, "Population," and Anonymous, "Inoculation," in *Encyclopédie, ou dictionnaire raisonné des sciences, des arts et des métiers, etc.*, ed. Denis Diderot and Jean le Rond d'Alembert, ARTFL Encyclopédie Project, ed. Robert Morrissey, University of Chicago, Spring 2013, http://encyclopedie.uchicago.edu/.

17. Samuel Auguste André David Tissot, *Inoculation Justifiée, ou Dissertation pratique et apologetique sur cette Méthode* (Lausanne: Chez Marc-Michel Bousquet & Co., 1754), 9.

18. Some key texts from the voluminous scholarship on the public sphere are Sarah Maza, *Private Lives and Public Affairs: The Causes Célèbres of Pre-revolutionary*

*France* (Berkeley: University of California Press, 1993); Keith Michael Baker, "Defining the Public Sphere in Eighteenth-Century France: Variations on a Theme by Habermas," in *Habermas and the Public Sphere*, ed. Craig Calhoun (Cambridge: MIT Press, 1991), 181–211; Robert Darnton, *The Forbidden Best-Sellers of Pre-revolutionary France* (New York: W. W. Norton, 1996), 169–246.

19. Tissot, *Inoculation Justifiée*, 153.

20. For a discussion of eighteenth-century quantitative techniques, see Rusnock, *Vital Accounts*; Andrea Rusnock, "The Weight of Evidence and the Burden of Authority: Case Histories, Medical Statistics, and Smallpox Inoculation," in *Medicine in the Enlightenment*, ed. Roy Porter (Amsterdam: Rodopi, 1995), 289–315; Marks, "When the State Counts Lives"; for a general account of the rise of quantitative methods, see Theodore M. Porter, *Trust in Numbers: The Pursuit of Objectivity in Science and Public Life* (Princeton, NJ: Princeton University Press, 1995).

21. L. Bradley, *Smallpox Inoculation: An Eighteenth Century Mathematical Controversy* (Nottingham, UK: Adult Education Department of the University of Nottingham, 1971), 10.

22. On smallpox inoculation and public health, see Meyer, "Divining the Pox." There was also a deep interest in preventative medicine in general. Vila, *Enlightenment and Pathology*, 43–107.

23. Jean le Rond d'Alembert, "Réflexions sur l'inoculation," in *Oeuvres Complètes* (Paris: A. Belin, 1821), 1:484.

24. Denis Diderot, "Sur deux mémoires de d'Alembert, l'un concernant le calcul des probabilités, l'autre l'inoculation," in *Oeuvres Complètes de Diderot*, ed. Jules Assézat (Nendeln, Liechtenstein: Kraus Reprint, 1875), 9:207.

25. On "society" as a grounding concept, see the discussion in chapter 1. See also Daniel Gordon, *Citizens without Sovereignty: Equality and Sociability in French Thought, 1670–1789* (Princeton, NJ: Princeton University Press, 1994), 51–85; Keith Baker, "Enlightenment and the Institution of Society: Notes for a Conceptual History," in *Main Trends in Cultural History: Ten Essays*, ed. William Melching and Wyger Velema (Amsterdam: Rodopi, 1994), 95–120; Sarah Maza, *The Myth of the French Bourgeoisie: An Essay on the Social Imaginary, 1750–1850* (Cambridge, MA: Harvard University Press, 2003), chap. 1.

26. Lynn Hunt, *The Family Romance of the French Revolution* (Berkeley: University of California Press, 1992), 17–40; Leslie Tuttle, "Celebrating the *Père de Famille*: Pronatalism and Fatherhood in Eighteenth-Century France," *Journal of Family History* 29, no. 4 (October 2004): 366–81. For the medical side of paternalism, see Lisa Smith, "The Relative Duties of Man: Domestic Medicine in England and France, ca. 1685–1740," *Journal of Family History* 33, no. 3 (July 2006): 237–56.

27. On the intersection of love and prudence, see Amanda Vickery, *The Gentleman's Daughter: Women's Lives in Georgian England* (New Haven, CT: Yale University Press, 2003), 39–86, 127–60.

28. Jeffrey Merrick, "Sexual Politics and Public Order in Late Eighteenth-Century France: The *Mémoires secrets* and the *Correspondance secrète*," *Journal of the History of Sexuality* 1, no. 1 (July 1990): 68–84.

29. Ibid., 81–84.

30. E.g., Charles Collé, *Journal et Mémoires de Charles Collé Sur les Hommes de Lettres, Les Ouvrages Dramatiques, et les Evénéments le plus Mémorables du Règne de Louis XV* (Paris: Librairie de Firmin Didot Frères, 1868), 2:48.

31. Charles-Marie de la Condamine, *Mémoires pour servir à l'histoire de l'inoculation de la Petite Vérole: Lûs à l'Académie Royale des Sciences en 1754, 1758 &1765*, (Paris: Imprimerie Royale, 1768), 1.

32. Ibid., 35.

33. On reason, emotion, and intellectual authority, see Jessica Riskin, *Science in the Age of Sensibility: The Sentimental Empiricists of the French Enlightenment* (Chicago: University of Chicago Press, 2002); Vila, *Enlightenment and Pathology*; E. C. Spary, *Utopia's Garden: French Natural History from Old Regime to Revolution* (Chicago: University of Chicago Press, 2000).

34. Tissot, *Inoculation Justifiée*, 154–55.

35. La Condamine, "Mémoires," 39.

36. Ibid., 42.

37. Ibid., 43.

38. Jean-Jacques Rousseau, *Emile, ou de l'éducation*, ed. Michel Launay (Paris: GF Flammarion, 1966), 165.

39. Riskin, *Science in the Age of Sensibility*, 1–2.

40. La Condamine, "Mémoires," 73–74.

41. Rusnock, "The Weight of Evidence and the Burden of Authority."

42. La Condamine, "Mémoires," 82.

43. Sara Stidstone Gronim, "Imagining Inoculation: Smallpox, the Body, and Social Relations of Healing in the Eighteenth Century," *Bulletin of the History of Medicine* 80, no. 2 (2006): 248–49.

44. Charles-Marie de la Condamine, *Lettres de M. de la Condamine à M. Daniel Bernoulli, Extrait du Mercure du mars et d'avril 1760*, p. 16, [stand-alone microfilm of extracts from *Mercure du Mars*, March and April 1760], Bibliothèque Nationale de France (hereafter cited as *Lettres de M. de la Condamine*).

45. Andrew Cantwell, *Dissertation sur l'Inoculation pour servir de réponse à celle de M. de la Condamine, de l'Académie Royale des Sciences, sur le même sujet* (Paris: Chez Delaguette, 1755).

46. La Condamine, *Lettres de M. de la Condamine*, 16.

47. Ibid., 17.

48. Jay M. Smith, *Nobility Reimagined: The Patriotic Nation in Eighteenth-Century France* (Ithaca, NY: Cornell University Press, 2005); Maza, *Myth of the French Bourgeoisie*; David A. Bell, *The Cult of the Nation in France: Inventing Nationalism, 1680–1800* (Cambridge, MA: Harvard University Press, 2003); John Shovlin, *The Political Economy of Virtue: Luxury, Patriotism, and the Origins of the French Revolution* (Ithaca, NY: Cornell University Press, 2006).

49. Claude-Louis-Michel de Sacy quoted in Smith, *Nobility Reimagined*, 175.

50. Brockliss and Jones, *Medical World*, 474–75; Vila, *Enlightenment and Pathology*, 63, 80–107.

51. Smith, *Nobility Reimagined*, 9–11, 175–80. See also Shovlin, *Political Economy of Virtue*.

52. Catriona Seth, *Les Rois Aussi en Mouraient: Les Lumières en lutte contre la petite vérole* (Paris: Les Editions Desjonquères, 2008), 114.

53. Lady Wortley Montagu to Mrs. S.C. [Miss Sarah Chiswell], 1 April 1717, in *Letters of the Right Honourable Lady Wortley Montagu*, 2:62–63.

54. Seth, *Les Rois Aussi en Mouraient*, 119.

55. Tissot, *Inoculation Justifiée*, 5.

56. La Condamine, "Mémoires," 4.

57. Elizabeth Fenn, *Pox Americana: The Great Smallpox Epidemic of 1775–82* (New York: Hill and Wang, 2001), 32.

58. Ibid., 36; Kenneth Silverman, *The Life and Times of Cotton Mather* (New York: Columbia University Press, 1985), 336–45.

59. Silverman, *Life and Times of Cotton Mather*, 347–48.

60. Julia Douthwaite, *The Wild Girl, the Natural Man, and the Monster: Dangerous Experiments in the Age of Enlightenment* (Chicago: University of Chicago Press, 2002), 72; Stuart Walker Strickland, "The Ideology of Self-Knowledge and the Practice of Self-Experimentation," *Eighteenth-Century Studies* 31, no. 4 (1998): 453–71; Simon Schaffer, "Self-Evidence," *Critical Enquiry* 18, no. 2 (1992): 327–62.

61. E.g., Louis-Pierre Manuel, *L'année françoise, ou vies des hommes qui ont honoré la France, ou par leurs talens, ou par leurs service et sur-tout par leurs vertus,* 4 vols. (Paris, 1789), 1:64–65; J. A. N. Condorcet, *Oeuvres de Condorcet*, ed. Arthur Condorcet-O'Connor (Paris: Firmin Didot Frères, 1847–1849), 2:201.

62. Tissot, *Inoculation Justifiée*, 8. See also James E. McClellan III, *Colonialism and Science: Saint Domingue in the Old Regime* (Baltimore, MD: Johns Hopkins University Press, 1992), 144; Londa Schiebinger, "Human Experimentation in the Eighteenth Century: Natural Boundaries and Valid Testing," in *The Moral Authority of Nature*, ed. Lorraine Daston and Fernando Vidal (Chicago: University of Chicago Press, 2003), 384–408.

63. Madame de Montbeillard, "Biographie de M. Gueneau de Montbeillard," in George-Louis LeClerc, comte de Buffon, *Correspondance inédite de Buffon*, ed. Henri Nadault de Buffon (Paris: L. Hachette et Cie, 1860), 1:344.

64. Comte de Buffon to Madame Gueneau de Montbeillard, 2 May 1766, in *Correspondance inédite de Buffon*, 1:103.

65. Madame de Montbeillard, "Biographie de M. Gueneau de Montbeillard," 1:342, 344.

66. Letter from Denis Diderot to Philibert Gueneau de Montbeillard, 27 November 1766 included as appendix in *Correspondance inédite de Buffon*, 1:343.

67. Manuel, *L'année françoise*, 4:274.

68. La Condamine, "Mémoires," 13–14.

69. Anonymous, "Inoculation," in *Encyclopédie* (italics in original).

70. Tronchin, "Inoculation," in *Encyclopédie* (italics in original).

71. La Condamine, "Mémoires," 13–14 (1755), 75–76 (1758).

72. Condorcet, *Oeuvres*, 3:659.

73. On Vaucanson, his ambitions, and his duck, see Jessica Riskin, "The Defecating Duck, or, The Ambiguous Origins of Artificial Life," *Critical Inquiry* 29, no. 4 (Summer 2003): 599–633.

74. Louise Dorothea von Meiningen, Duchesse de Saxe-Gotha, to Voltaire, 28 April 1759, accessible via Electronic Enlightenment, http://dx.doi.org/10.13051/ee:doc/voltfrVF1040136a1c (accessed 17 September 2015).

75. La Condamine, "Mémoires," 82.

76. Ibid.

77. For another savant testing inoculation on himself, see Buffon's comments on the chevalier de Chastellux. Buffon, *Correspondance inédite de Buffon*, 1:343.

78. Condorcet, *Oeuvres*, 3:195–96.

79. This was true of the reading public in general. Seth, *Les Rois Aussi en Mouraient*, 137.

80. Maza, *Private Lives and Public Affairs*; Russo, *Styles of Enlightenment*.

81. Cantwell, *Dissertation sur l'Inoculation*, 2. On the previous page, he had also noted: "La seule passion qui m'anime est l'intérêt que tout bon Citoyen prend à l'avantage commun de l'humanité; & j'y suis obligé plus qu'un autre, en qualité de Médecin." This, too, echoed pro-inoculation writers' claims to civic virtue.

82. *Lettres de M. de la Condamine*, 4–5.

83. Ibid., 50–51.

CHAPTER FOUR

1. J. A. N. Condorcet, *Oeuvres de Condorcet*, ed. Arthur Condorcet-O'Connor (Paris: Firmin Didot Frères, 1847–1849), 2:453.

2. Julia Douthwaite, *The Wild Girl, the Natural Man, and the Monster: Dangerous Experiments in the Age of Enlightenment* (Chicago: University of Chicago Press, 2002); Anne C. Vila, *Enlightenment and Pathology: Sensibility in the Literature and Medicine of Eighteenth-Century France* (Baltimore, MD: Johns Hopkins University Press, 1998); Jean Bloch, *Rousseauism and Education in Eighteenth-Century France* (Oxford: Society on Voltaire and the Eighteenth Century, 1995).

3. Kate Retford, *The Art of Domestic Life: Family Portraiture in Eighteenth-Century England* (New Haven, CT: Yale University Press, 2006), 83–97, 115–32, 145–47; Lynn Hunt, *The Family Romance of the French Revolution* (Berkeley: University of California Press, 1992), 17–40; Leslie Tuttle, "Celebrating the *Père de Famille*: Pronatalism and Fatherhood in Eighteenth-Century France," *Journal of Family History* 29, no. 4 (October 2004): 366–81; Jeffrey Merrick, "Sexual Politics and Public Order in Late Eighteenth-Century France: The *Mémoires secrets* and the *Correspondance secrète*," *Journal of the History of Sexuality* 1, no. 1 (July 1990): 68–84.

4. Natasha Gill, *Educational Philosophy in the French Enlightenment: From Nature to Second Nature* (Farnham, UK: Ashgate University Press, 2010), 39.

5. Retford, *Art of Domestic Life*, 131–32.

6. Gill, *Educational Philosophy*, 1–4. Although some philosophes adopted a more fixed understanding of human nature, many believed in varying degrees of malleability. Michael E. Winston, *From Perfectibility to Perversion: Meliorism in Eighteenth-Century France* (New York: Peter Lang, 2005), 20–37.

7. Marilyn Francus, *Monstrous Motherhood: 18th-Century Culture and the Ideology of Domesticity* (Baltimore, MD: Johns Hopkins University Press, 2012), 71–73.

8. Jennifer Popiel, *Rousseau's Daughters: Domesticity, Education, and Autonomy in Modern France* (Lebanon, NH: University Press of New England, 2008), 56–60, 96–98; on lack of physical activity as degenerative in later childhood, see Sean Quinlan, *The Great Nation in Decline: Sex, Modernity, and Health Crises in Revolutionary France c. 1750–1850* (Abindon: Ashgate University Press, 2007), 26–30.

9. Dale Van Kley, *The Jansenists and the Expulsion of the Jesuits from France, 1757–1765* (New Haven, CT: Yale University Press, 1975).

10. Charles R. Bailey, "The French Clergy and the Removal of Jesuits from Secondary Schools, 1761–1762," *Church History* 48, no. 3 (September 1979): 305–19, esp. 308–12.

11. Van Kley, *The Jansenists and the Expulsion of the Jesuits from France*, 148.

12. A. Lynn Martin, *The Jesuit Mind: The Mentality of an Elite in Early Modern France* (Ithaca, NY: Cornell University Press, 1988); Aldo Scaglione, *The Liberal Arts and the Jesuit College System* (Philadelphia: John Benjamins, 1986).

13. Jean Le Rond d'Alembert and Edme-François Mallet. "College [abridged]," in *The Encyclopedia of Diderot & d'Alembert Collaborative Translation Project*, trans. Nelly S. Hoyt and Thomas Cassirer (Ann Arbor: Michigan Publishing, University of Michigan Library, 2003), http://hdl.handle.net/2027/spo.did2222.0000.144 (accessed 5 August 2013). Originally published as "Collège [abridged]," in *Encyclopédie ou Dictionnaire raisonné des sciences, des arts et des métiers* (Paris, 1753), 3:634–37.

14. Harvey Chisick, *The Limits of Reform in the Enlightenment: Attitudes toward the Education of the Lower Classes in Eighteenth-Century France* (Princeton, NJ: Princeton University Press, 1981); Popiel, *Rousseau's Daughters*, 34–36.

15. Popiel, *Rousseau's Daughters*, 27–36.

16. Jean Ranson quoted in Robert Darnton, *The Great Cat Massacre and Other Episodes in French Cultural History* (New York: Vintage Books, 1985), 236; for more on the popularity of education, see Jeremy Caradonna's discussion of essay contests debating the topic in *The Enlightenment in Practice: Academic Prize Contests and Intellectual Culture in France, 1670–1794* (Ithaca, NY: Cornell University Press, 2012), 112–16.

17. Douthwaite, *The Wild Girl, the Natural Man, and the Monster*, 137–41.

18. César Chesneau Du Marsais, "Education," in *The Encyclopedia of Diderot & d'Alembert Collaborative Translation Project*, trans. Carolina Armenteros (Ann Arbor: Michigan Publishing, University of Michigan Library, 2007), http://hdl.handle.net/2027/spo.did2222.0000.390 (accessed 5 August 2013). Originally published as "Education," in *Encyclopédie ou Dictionnaire raisonné des sciences, des arts et des métiers* (Paris, 1755), 5:397–403.

19. R. R. Palmer, *The Improvement of Humanity: Education and the French Revolution* (Princeton, NJ: Princeton University Press, 1985), 56.

20. Gill, *Educational Philosophy*, 94.

21. B. d'Andlau, *La Jeunesse de Madame de Staël (de 1766 à 1786) avec des documents inédits* (Paris-Genève: Librairie Droz, 1970), 26.

22. Andlau, *Jeunesse de Madame de Staël*, 7; Madelyn Gutwirth, "Suzanne Necker's Legacy: Breastfeeding as Metonym in Germaine de Staël's *Delphine*," *Eighteenth-Century Life* 28, no. 2 (Spring 2004): 17–40, 27.

23. Popiel, *Rousseau's Daughters*, 110.

24. Carolyn Purnell, "Instruments Endowed with Sensibility: Remaking Society

through the Body in Eighteenth-Century France," (PhD diss., University of Chicago, 2013), 189.

25. Popiel, *Rousseau's Daughters*, 14; on motherhood and virtue, see Lesley Walker, *A Mother's Love: Crafting Feminine Virtue in Enlightenment France* (Lewisburg, PA: Bucknell University Press, 2008), 37–68; Retford, *Art of Domestic Life*, 83–97.

26. Lieselotte Steinbrügge, *The Moral Sex: Women's Nature in the French Enlightenment*, trans. Pamela E. Selwyn (New York: Oxford University Press, 1995), 62; Popiel, *Rousseau's Daughters*, 47; see also Adrian O'Connor, "Nature, Nurture, and the Social Order: Imagining Lessons and Lives for Women in Ancien Régime France," *French Politics, Culture, and Society* 30, no. 1 (Spring 2012): 2–4, 12–15.

27. Walker, *A Mother's Love*, 36.

28. Purnell, "Instruments Endowed with Sensibility," 174, 213.

29. If they did not have children of their own, they invented fictional relatives to advise. Nadine Bérenguier, *Conduct Books for Girls in Enlightenment France* (Farnham, UK: Ashgate University Press, 2011), 27–29.

30. Émilie Du Châtelet to Thierrot, 23 December 1737, *Les Lettres de la Marquise du Châtelet*, ed. Theodore Besterman (Geneva: Institut et Musée Voltaire, 1958), 1:202; although Du Châtelet searched for tutors, she was not satisfied by those she found and directed her son's education for at least three years. Judith P. Zinsser, *La Dame d'Esprit: A Biography of the Marquise Du Châtelet* (New York: Viking, 2006), 42.

31. Lorraine Daston and Katherine Park, *Wonders and the Order of Nature, 1150–1750* (New York: Zone Books, 2001), 354.

32. For a quick overview of Du Châtelet's theoretical foundations, see Patrick Guyot, "La pédagogie des *Institutions de physique*," in *Émilie Du Châtelet: Éclairages et documents nouveaux*, ed. Ulla Kölving and Olivier Courcelle (Ferney-Voltaire: Centre International d'Étude du XVIIIe Siècle, 2008), 267.

33. Émilie Du Châtelet, *Institutions de Physique* (Paris: Chez Prault fils, 1740), 1, 3, 2, 7.

34. Mark Motley, *Becoming a French Aristocrat: The Education of the Court Nobility, 1580–1715* (Princeton, NJ: Princeton University Press, 1990), 106–7.

35. Jay Smith, *The Culture of Merit: Nobility, Royal Service, and the Making of Absolute Monarchy in France, 1600–1789* (Ann Arbor: University of Michigan Press, 1996), 66–67; Motley, *Becoming a French Aristocrat*.

36. Émilie Du Châtelet, "Institutions de Physique," Nouvelles Acquisitions Françaises, MS 12265, Bibliothèque Nationale de France, site Richelieu, Paris, France.

37. Smith, *Culture of Merit*, 20–21.

38. Jay M. Smith, *Nobility Reimagined: The Patriotic Nation in Eighteenth-Century France* (Ithaca, NY: Cornell University Press, 2005), 186–205.

39. On nobility and military reform, see Smith, *Nobility Reimagined*, 198–205.

40. Ken Alder, "Stepson of the Enlightenment: Duc du Châtelet, the Colonel Who 'Caused' the French Revolution," *Eighteenth-Century Studies* 32, no. 1 (1998): 1–18, 8.

41. Alder, "Stepson of the Enlightenment," 4–5.

42. Émilie du Châtelet to Frederick the Great, 25 April 1740, *Les Lettres de la Marquise du Châtelet*, 2:13–14.

43. Du Châtelet, *Institutions*, 5.

44. Émilie du Châtelet to Johann Bernoulli, 30 May 1744, *Les Lettres de la Marquise du Châtelet*, 2:116.

45. John R. Iverson, "A female member of the Republic of Letters: Du Châtelet's Portrait in the *Bilder-Sal [ . . . ] brümhter Schrifftsteller*," in *Émilie Du Châtelet: Rewriting Enlightenment Philosophy and Science*, ed. Judith P. Zinsser and Julie Chandler Hays (Oxford: Voltaire Foundation, 2006), 35–51, esp. 41–45. For more on Du Châtelet's reputation, see Massimo Mazzotti, "Mme Du Châtelet, Academicienne de Bologne," in Kölving and Courcelle, *Émilie Du Châtelet*, 123–25.

46. Mary Terrall, "Émilie Du du Châtelet and the Gendering of Science," *History of Science* 33 (1995): 303.

47. Ibid., 293.

48. Zinsser, *La Dame d'Esprit*, 40–41.

49. Ibid., 221.

50. Marisa Linton, "Virtue Rewarded? Women and the Politics of Virtue in Eighteenth-Century France, Part II," *History of European Ideas* 26, no. 1 (2000): 51–65.

51. Religion was generally the foundation of girls' education. Martine Sonnet, *L'éducation des filles au temps des Lumières* (Paris: Les Editions du Cerf, 1987), 234–40.

52. Denis Diderot to Sophie Volland, 13 July 1762, in *Correspondance*, ed. George Roth (Paris, 1955–1970), 4:86.

53. Ibid.

54. Denis Diderot to Sophie Volland, 16 September 1762, in *Correspondance*, 4:154.

55. Denis Diderot to the Volland women, 24 July 1769, in *Correspondance*, 9:84; Denis Diderot to Madame de Maux, September 1769, in *Correspondance*, 9:130; see also Carol Blum, *Diderot: The Virtue of a Philosopher* (New York: Viking, 1974), 102.

56. Diderot, *"Rameau's Nephew" and "D'Alembert's Dream,"* ed. and trans. Leonard Tancock (New York: Penguin Books, 1966), 56–57. Published posthumously, but seemingly composed c. 1761. L. W. Tancock, introduction to ibid., 23.

57. Denis Diderot to the Volland women, 24 July 1769, in *Correspondance*, 9:84.

58. Denis Diderot to Melchior Grimm, 12 November 1769, in *Correspondance*, 9:206.

59. Denis Diderot to Jacques-André Naigeon, April 1772, in *Correspondance*, 12:47.

60. Denis Diderot to Sophie Volland, 22 November 1768, in *Correspondance*, 8:232.

61. See also Blum, *Diderot*, 103–4. For a discussion of Diderot's sexual politics in general, see Mary Trouille, "Sexual/Textual Politics in the Enlightenment: Diderot and d'Epinay Respond to Thomas's Essay on Women," *Romanic Review* 84, no. 2 (March 1994): 98–116.

62. Denis Diderot to Sophie Volland, 28 August 1768, in *Correspondance*, 8:103.

63. Denis Diderot to Sophie Volland, 22 November 1768, in *Correspondance*, 8:231.

64. *The Nun* was too controversial to risk printing; it was only published posthumously.

65. Denis Diderot, *The Nun*, trans. Leonard Tancock (New York: Penguin Books, 1972), 136.

66. Denis Diderot to Melchior Grimm, March 1771, and to John Wilkes, 19 October 1771, in *Correspondance*, 10:245, 11:211; for more on Biheron, see Londa Schiebinger, *The Mind Has No Sex? Women in the Origins of Modern Science* (Cambridge, MA: Harvard University Press, 1991), 27–30.

67. Denis Diderot and Pierre Tarin, "Anatomie," in *Encyclopédie, ou dictionnaire raisonné des sciences, des arts et des métiers, etc.*, ed. Denis Diderot and Jean le Rond d'Alembert, 1:409, ARTFL Encyclopédie Project, ed. Robert Morrissey, University of Chicago, Spring 2013, http://encyclopedie.uchicago.edu/.

68. Rebecca Messbarger, *The Lady Anatomist: The Life and Work of Anna Morandi Manzolini* (Chicago: University of Chicago Press, 2010), 21–23; Paula Findlen, "Science as a Career in Enlightenment Italy: The Strategies of Laura Bassi," *Isis* 84, no. 3 (September 1993): 441–69, 452–53; Paula Findlen, "Anatomy Theaters, Botanical Gardens, and Natural History Collections," in *The Cambridge History of Science*, vol. 3, *Early Modern Science*, ed. Katharine Park and Lorraine Daston (Cambridge: Cambridge University Press, 2006), 272–89, 274–80.

69. Shane Agin, "Sex Education in the Enlightened Nation," *Studies in Eighteenth-Century Culture* 37 (2007): 1–21. On other "scientific" methods to provide sexual education and thereby guard sexual virtue, see Bérenguier, *Conduct Books for Girls*, 46–47.

70. On doctors, see Diderot and Tarin, "Anatomie," 1:410.

71. Denis Diderot, *Mémoires pour Catherine II* (Paris: Éditions Garnier Frères, 1966), 87.

72. Denis Diderot to l'abbé Diderot, 13 November 1772, in *Correspondance*, 12:164.

73. Denis Diderot to John Wilkes, 19 October 1771, in *Correspondance*, 11:211.

74. Diderot, *Mémoires*, 86. Although Diderot strongly suggested that anatomy lessons covered sexual reproduction, in his memoir to Catherine, Diderot claimed that only married women learned about reproductive organs in these anatomy lessons. Ibid., 88.

75. Ibid., 90.

76. William F. Edmiston, *Diderot and the Family: A Conflict of Nature and Law* (Palo Alto, CA: Anma Libri, 1985), 86–117.

77. Denis Diderot to Denise Diderot, 23 March 1770, in *Correspondance*, 10:30–31.

78. For comparative perspective, see Dena Goodman, "Marriage Choice and Marital Success: Reasoning about Marriage, Love, and Happiness," in *Family and State in Early Modern France*, ed. Suzanne Desan and Jeffrey D. Merricks (University Park: Penn State University Press, 2009), 26–61.

79. Denis Diderot to Denise Diderot, 4 October 1770, in *Correspondance*, 10:132. Dena Goodman has noted a similar flexibility in women's understandings of marriage. Goodman, "Marriage Choice and Marital Success." See also Amanda Vickery, *The Gentleman's Daughter: Women's Lives in Georgian England* (New Haven, CT: Yale University Press, 2003), 39–86, 127–60.

80. Denis Diderot to Denise Diderot, 27 August 1771, in *Correspondance*, 11:138.

81. Denis Diderot to Melchior Grimm, 9 December 1772, in *Correspondance*, 12:178–79.

82. Marisa Linton, "Virtue Rewarded? Women and the Politics of Virtue in Eighteenth-Century France, Part I," *History of European Ideas* 26, no. 1 (2000): 35–49; and Linton, "Virtue Rewarded? . . . Part II." On luxury, morality, and women, see Sarah Maza, *The Myth of the French Bourgeoisie: An Essay on the Social Imaginary, 1750–1850* (Cambridge, MA: Harvard University Press, 2003), chap. 2; Jennifer M. Jones, *Sexing la Mode: Gender, Fashion and Commercial Culture in Old Regime France* (Oxford: Berg, 2004).

83. Denis Diderot to Sophie Volland, 22 November 1768, in *Correspondance*, 8:231.

84. Denis Diderot, "Essai sur les femmes," in *Qu'est-ce qu'une femme?*, ed. Elisabeth Badinter (Paris: P.O.L., 1989), 180. A similar paragraph appears in Diderot, *Mémoires*, 91; he published this text in 1773, a year after "Sur les femmes."

85. Diderot, "Essai sur les femmes," 174. "Sur les femmes" was not wholly pessimistic. Diderot praised women's sensitivity and the value of their "commerce" with men of letters. Ibid., 166 and 184–85.

86. Rosena Davison, "Happy Marriage: Myth or Reality in Eighteenth-Century France? The Case of Madame d'Épinay and Her Family," *Dalhousie French Studies* 56 (Fall 2001): 116–24, esp. 119–20.

87. Rousseau had lived on D'Epinay's estate but did not see himself as in her debt. When she had to leave for Geneva because of health issues, Rousseau opted not to accompany her; many of the philosophes thought he owed her as much. The situation gradually escalated until Rousseau's break with the philosophes and—more devastatingly—with his longtime friend Diderot.

88. Rosena Davison, "Madame D'Epinay's Contribution to Girls' Education," in *Femmes Savantes et Femmes d'Esprit: Women Intellectuals of the French Eighteenth Century*, ed. Roland Bonnel and Catherine Rubinger (New York: Peter Lang, 1994), 220–24.

89. "Letter to My Daughter's Governess" appeared in the *Correspondence Littéraire* in 1756 and again in her first short book, *Les Moments Heureux*, published in 1758.

90. Louise d'Epinay, "Letter to My Daughter's Governess," appendix to Louise d'Epinay, *Lettres à mon fils: Essais sur l'éducation*, ed. Ruth Plaut Weinreb (Sandwich, MA: Wayside Publishing, 1989), 113.

91. For more on the moral value of motherhood, see Popiel, *Rousseau's Daughters*.

92. Ruth Plaut Weinreb, *Eagle in a Gauze Cage: Louise d'Epinay, Femme de Lettres* (New York: AMS Press, 1993), 75.

93. Louise D'Epinay quoted in ibid. Louise D'Epinay to Ferdinando Galiani, 15 July 1770, in *Correspondance*, vol. 1, *1769–1770*, ed. Georges Dulac and Daniel Maggetti (Paris: Les Editions Desjonquères, 1992), 209.

94. Ferdinando Galiani to Louise D'Epinay, 19 January 1771, in *Correspondance*, vol. 2, *1771–fevrier 1772*, ed. Georges Dulac and Daniel Maggetti (Paris: Les Editions Desjonquères, 1993), 35–36. Even while praising her *Conversations d'Emilie*, he once again reminded her that education was a capricious endeavor. Francis Steegmuller, *A Woman, A Man, and Two Kingdoms: The Story of Madame d'Épinay and the Abbé Galiani* (New York: Alfred A. Knopf, 1991), 200.

95. Louise D'Epinay to Ferdinando Galiani, 2 September 1770, in *Correspondance*, 1:248.

96. Louise D'Epinay, *Les Conversations d'Emilie*, ed. Rosena Davison (Oxford: Voltaire Foundation, 1996), 48.

97. A common strategy for pedagogical writers. Bérenguier, *Conduct Books for Girls*, 26–27.

98. D'Epinay, *Conversations*, 48, 23.

99. Anne Schroder, "Going Public against the Academy in 1784: Mme de Genlis Speaks Out on Gender Bias," *Eighteenth-Century Studies* 32, no. 3 (1999): 376–82, 377.

100. D'Epinay, *Conversations*, 22.

101. I borrow this phrase from Jessica Riskin, *Science in the Age of Sensibility: The Sentimental Empiricists of the French Enlightenment* (Chicago: University of Chicago Press, 2002).

102. D'Epinay, *Conversations*, 50.

103. Ibid., 47.

104. D'Epinay, *Lettres à mon fils*, 57.

105. D'Epinay, *Conversations*, 69.

106. Ibid., 51–52.

107. Ibid., 62.

108. Guillaume Ansart, "Condorcet, Social Mathematics, and Women's Rights," *Eighteenth-Century Studies* 42, no. 3 (Spring 2009): 347–62; Madelyn Gutwirth, "Civil Rights and the Wrongs of Women," in *A New History of French Literature*, ed. Denis Hollier (Cambridge, MA: Harvard University Press, 1989), 558–66; Joan B. Landes, *Women and the Public Sphere in the Age of the French Revolution* (Ithaca, NY: Cornell University Press, 1988), 112–17.

109. Rebecca Rogers, *From the Salon to the Schoolroom: Educating Bourgeois Girls in Nineteenth-Century France* (University Park, PA: Penn State University Press, 2005), 21. He did, however, eliminate girls' education from the upper levels of this public education, presumably to make the plan politically viable. Palmer, *Improvement of Humanity*, 124–25.

110. Keith Baker, *Condorcet: From Natural Philosophy to Social Mathematics* (Chicago: University of Chicago Press, 1975), 293–96; Charles Coulston Gillispie, *Science and Polity in France: The Revolutionary and Napoleonic Years* (Princeton, NJ: Princeton University Press, 2004), 110–14.

111. Winston, *From Perfectibility to Perversion*, 123–30; for an overview of the Condorcet plan, see Palmer, *Improvement of Humanity*, 124–37.

112. Marquis de Condorcet, "Avis d'un proscrit à sa fille," Correspondance de la famille O'Connor, MS 2475, Bibliothèque de l'Institut, Paris, France.

113. As a result, Condorcet restricted himself to the vague advice that she should choose "un genre de travail où la main ne soit pas occupée seule, où l'esprit s'exerce sans trop de fatigue; un travail qui dédommage de ce qu'il coûte par le plaisir qu'il procure" (ibid.).

114. Ibid.

115. Ibid.

116. On the potential backlash to an overly assertive declaration of paternal excellence, see Retford, *Art of Domestic Life*, 125.

117. Walker, *A Mother's Love*, 134–63.

CHAPTER FIVE

1. Antoine Lavoisier, "Résultats de Quelques Expériences d'Agriculture et Réflexions sur leurs Relations avec l'Economie Politique, Lu en 1788 à la Société d'agriculture de Paris," *Annales de Chimie* 25 (December 1792): 822.

2. Jean-Pierre Poirier, *Lavoisier: Chemist, Biologist, Economist*, trans. Rebecca Balinski (Philadelphia: University of Pennsylvania Press, 1996), 121; Madison Smart

Bell, *Lavoisier in the Year One: The Birth of a New Science in an Age of Revolution* (New York: W. W. Norton, 2005), 6.

3. Marie-Anne Lavoisier, "Notice Biographique de Lavoisier," ed. Charles Gillispie, *Revue d'histoire des sciences et de leurs applications* 9, no. 1 (1956): 52–61, 59.

4. André Cauderon, "Lavoisier et l'agronomie," and Maurice Gobillon, "La formation du grand domaine de Freschines par Lavoisier," in *Il y a 200 ans Lavoisier: Actes du Colloque organisé à l'occasion du bicentenaire de la mort d'Antoine Laurent Lavoisier, le 8 mai 1794*, ed. Christine Demeulenaere-Douyère (London: Tec & Doc, 1994), 19–28 and 235–48.

5. John Shovlin, *The Political Economy of Virtue: Luxury, Patriotism, and the Origins of the French Revolution* (Ithaca, NY: Cornell University Press, 2006), 80–117; Elizabeth Fox-Genovese, *The Origins of Physiocracy: Economic Revolution and Social Order in Eighteenth-Century France* (Ithaca, NY: Cornell University Press, 1976), 218–20; Philip T. Hoffman, *Growth in a Traditional Society: The French Countryside, 1450–1615* (Princeton, NJ: Princeton University Press, 1996), 165–66; Liana Vardi, *The Physiocrats and the World of the Enlightenment* (Cambridge: Cambridge University Press, 2012), 141.

6. François de Fenelon, *Telemachus, Son of Ulysses*, ed. and trans. Patrick Riley (Cambridge: Cambridge University Press, 1994), 166–67, 168.

7. Liana Vardi, *The Land and the Loom: Peasants and Profit in Northern France, 1680–1800* (Durham, NC: Duke University Press, 1993), 87–109; Hoffman, *Growth in a Traditional Society*, 81–142.

8. G. E. Mingay, introduction to *The Agricultural Revolution: Changes in Agriculture, 1650–1880*, ed. G. E. Mingay (London: Adam & Charles Black, 1977), 19–23; Richard Drayton, *Nature's Government: Science, Imperial Britain, and the 'Improvement' of the World* (New Haven, CT: Yale University Press, 2000), 85–94.

9. J. V. Beckett, *The Agricultural Revolution* (Oxford: Basil Blackwell, 1990), 16–25.

10. Marie-Anne Lavoisier, "Notice Biographique," 59.

11. Shovlin, *Political Economy of Virtue*, 83.

12. Ibid., 80–117.

13. Sarah Maza, *The Myth of the French Bourgeoisie: An Essay on the Social Imaginary, 1750–1850* (Cambridge, MA: Harvard University Press, 2003); Jay M. Smith, *Nobility Reimagined: The Patriotic Nation in Eighteenth-Century France* (Ithaca, NY: Cornell University Press, 2005); Marisa Linton, *The Politics of Virtue in Enlightenment France* (New York: Palgrave, 2001); David A. Bell, *The Cult of the Nation: Inventing Nationalism, 1680–1800* (Cambridge, MA: Harvard University Press, 2001); Johnson Kent Wright, *A Classical Republican in Eighteenth-Century France: The Political Thought of Mably* (Palo Alto, CA: Stanford University Press, 1997).

14. Jay M. Smith, *Monsters of the Gévaudan: The Making of a Beast* (Cambridge, MA: Harvard University Press, 2011), 60–106.

15. Shovlin, *Political Economy of Virtue*, 5–7.

16. Jean-Jacques Rousseau, *Discours sur les origines et les fondements de l'inégalité parmi les hommes* (Paris: GF Flammarion, 2011).

17. Maza, *Myth of the French Bourgeoisie*, 41–68.

18. Shovlin, *Political Economy of Virtue*, 10.

19. Sarah Maza, *Private Lives and Public Affairs: The Causes Célèbres of Pre-revolutionary France* (Berkeley: University of California Press, 1993), 68–85; Vardi, *The Land and the Loom*, 234–40; on elite fascination with peasants in general, see Amy S. Wyngaard, *From Savage to Citizen: The Invention of the Peasant in the French Enlightenment* (Newark: University of Delaware Press, 2004).

20. Henri-Louis Duhamel du Monceau, *École d'Agriculture* (Paris: chez les frères Estienne, 1759), 5–6.

21. Cauderon, "Lavoisier et l'agronomie"; and Maurice Gobillon, "Lavoisier, son domaine de Freschines et l'agronomie," in Demeulenaere-Douyère, *Il y a 200 ans Lavoisier*; Jean-Pierre Poirier, *Lavoisier: Chemist, Biologist, Economist*, trans. Rebecca Balinski (Philadelphia: University of Pennsylvania Press, 1996), 121–31, 198–216.

22. Shovlin, *Political Economy of Virtue*, 162.

23. Antoine Lavoisier, "Résultats," 813. These "résultats" were first read to the Société d'Agriculture in 1788 and were printed later.

24. Shovlin, *Political Economy of Virtue*, 92.

25. Antoine Lavoisier, "Résultats," 816.

26. On virtue, see Smith, *Nobility Reimagined*; Linton, *Politics of Virtue in Enlightenment France*.

27. Pierre de Saint-Jacob, *Les Paysans de la Bourgogne du Nord au Dernier Siècle de l'Ancien Régime* (Paris: Société les Belles Lettres, 1960), 339, 397.

28. Antoine Lavoisier, "Résultats," 813–14.

29. Bernadette Bensaude-Vincent, "The Balance: Between Chemistry and Politics," *The Eighteenth Century* 33, no. 2 (Fall 1992): 217–37, 226.

30. Marie-Anne Lavoisier, "Notice Biographique," 59.

31. Antoine Lavoisier, "Résultats," 814.

32. Ibid.

33. Antoine Lavoisier to Lubert, 31 July 1785, in *Oeuvres de Lavoisier: Correspondance*, ed. René Fric (Paris: Michel Albion, 1957), 4:143.

34. Marie-Anne Lavoisier, "Notice Biographique," 59.

35. Jessica Riskin, "The 'Spirit of System' and the Fortunes of Physiocracy," *History of Political Economy* 35 (2003): 42–73, 53; Poirier, *Lavoisier*, 247–73.

36. Jessica Riskin, *Science in the Age of Sensibility: The Sentimental Empiricists of the French Enlightenment* (Chicago: University of Chicago Press, 2002), 106.

37. Shovlin, *Political Economy of Virtue*, 103.

38. Fox-Genovese, *Origins of Physiocracy*, 126; Christopher Hodson, *The Acadian Diaspora: An Eighteenth-Century History* (New York: Oxford University Press, 2012), 178–80.

39. Antoine Lavoisier, "Résultats," 819.

40. Michael Kwass, *Privilege and the Politics of Taxation in Eighteenth-Century France* (Cambridge: Cambridge University Press, 2006); J. F. Bosher, *French Finances, 1770–1795: From Business to Bureaucracy* (Cambridge: Cambridge University Press, 1970); James C. Riley, *The Seven Years' War and the Old Regime in France: The Economic and Financial Toll* (Princeton, NJ: Princeton University Press, 1996).

41. Kwass, *Privilege and the Politics of Taxation*.

42. For example: "Le véritable restaurateur de l'agriculture sera le ministre qui

forcera ses successeurs de s'en occuper, en plaçant dans leurs fonctions annuelles celle d'employer avec utilité les fonds qu'il aura formés pour son encouragement: ce sera vous." Tillet, Poissonnier, Lavoisier, d'Arcet, et Du Pont to Controller General of Finances, 15 February 1786, in *Oeuvres de Lavoisier*, 4:203–4.

43. Riskin, "The 'Spirit of System' and the Fortunes of Physiocracy," 53.

44. Riskin, *Science in the Age of Sensibility*.

45. Robert Forster, *The Nobility of Toulouse in the Eighteenth Century* (Baltimore, MD: Johns Hopkins University Press, 1960); John Shovlin, "Political Economy and the French Nobility," in *The French Nobility in the Eighteenth Century: Reassessments and New Approaches*, ed. Jay M. Smith (University Park: Penn State University Press, 2006), 124–35. On merit and nobility more generally, see Jay M. Smith, *The Culture of Merit: Nobility, Royal Service, and the Making of Absolute Monarchy in France, 1600–1789* (Ann Arbor: University of Michigan Press, 1996); Smith, *Nobility Reimagined*, 104–42; Maza, *Myth of the French Bourgeoisie*, 27–36.

46. Marie-Anne Lavoisier, "Notice Biographique," 60.

47. In his "Fragments d'un éloge de Colbert," Lavoisier noted (seemingly with approval) that the minister worked in "le silence du cabinet." Antoine-Laurent Lavoisier, "Fragments d'un Eloge de Colbert" (1771), in *Oeuvres de Lavoisier*, 6:110, 114; Bensaude-Vincent, "The Balance."

48. Antoine Lavoisier to Lubert, 5 August 1785 and 31 July 1785, in *Oeuvres de Lavoisier*, 4:144, 143.

49. Antoine Lavoisier to Lubert, 5 August 1785, in *Oeuvres de Lavoisier*, 4:145.

50. Antoine Lavoisier to municipal officers of Blois, 16 March 1789, in *Oeuvres de Lavoisier*, 6:33–34.

51. Antoine Lavoisier, "Résultats," 822.

52. Marie-Anne Lavoisier, "Notes sur Lavoisier," Éloges des savants, Papiers et correspondance du baron George Cuvier, MS 3145, Bibliothèque de l'Institut, Paris, France.

53. Ibid.

54. Smith, *Nobility Reimagined*, 143–81.

55. Marie-Anne Lavoisier, "Notes sur Lavoisier."

56. Poirier, *Lavoisier*, 124.

57. Fourcroy, *Notice sur la Vie et les Travaux de Lavoisier, lue le 15 Thermidor, an 4, au Lycée des Arts* (Paris: Imprimerie de la Feuille du Cultivateur, 1796), 42.

58. Ibid., 43.

59. Eulogy, by Pierre Samuel Du Pont de Nemours, quoted in Vardi, *Physiocrats*, 189.

60. Voltaire to Nicolas-Claude Thieriot, 17 September 1759, in *Correspondance de Voltaire*, ed. Theodore Besterman (Paris: Editions Gallimard, 1980), 5:608.

61. Ibid.

62. Louis-Pierre Manuel, *L'année françoise, ou vies des hommes qui ont honoré la France, ou par leurs talens, ou par leurs service et sur-tout par leurs vertus*, 4 vols. (Paris, 1789), 1:241.

63. John C. O'Neal, "Morality in Rousseau's Public and Private Society at Clarens," *Revue de Métaphysique et de Morale* 89, no. 1 (January–March 1984): 58–67.

64. Jean-Jacques Rousseau, *Julie, Or the New Heloïse: Letters of Two Lovers Who Live in a Small Town at the Foot of the Alps*, ed. and trans. Jean Vaché and Philip

Stewart (Hanover, NH: Dartmouth University Press/University Press of New England, 1997), 364.

65. Ibid., 365.

66. Ibid., 372.

67. Ibid., 365. See also Michael E. Winston, *From Perfectibility to Perversion: Meliorism in Eighteenth-Century France* (New York: Peter Land, 2005), 60–62.

68. Sarah Maza, *Servants and Masters in Eighteenth-Century France: The Uses of Loyalty* (Princeton, NJ: Princeton University Press, 1983), 300–301.

69. Marie-Anne Lavoisier, "Notes sur Lavoisier"; Rousseau, *Julie, Or the New Heloïse*, 365.

70. "La Nouvelle Héloïse," Fonds Antoine-Laurent Lavoisier, MS 1711, Archives de l'Académie des Sciences, Paris, France.

71. On *bienfaisance* and natural virtue, see Linton, *Politics of Virtue in Enlightenment France*, 67–74, 184–86.

72. Emma Rothschild, *Economic Sentiments: Adam Smith, Condorcet, and the Enlightenment* (Cambridge, MA: Harvard University Press, 2001), 1, 28, and passim.

73. Antoine Lavoisier to [Gravier de Vergennes], 17 June 1785, in *Oeuvres de Lavoisier*, 4:130.

74. Marie-Anne Lavoisier, "Notes sur Lavoisier."

75. Antoine Lavoisier, "Résultats," 819.

76. Ibid.

77. Louis-Michel Lefebvre to Marie-Anne Lavoisier, 16 July 1784, in *Oeuvres de Lavoisier*, 4:24–25.

78. Louis-Michel Lefebvre to Marie-Anne Lavoisier, 11 Feburary 1790, in *Oeuvres de Lavoisier*, 6:112–3.

79. On peasants as "backward," see Vardi, *The Land and the Loom*, 87–109.

80. Denis Diderot, "Farm Laborer," in *The Encyclopedia of Diderot & d'Alembert Collaborative Translation Project*, trans. Stephen J. Gendzier (Ann Arbor: Michigan Publishing, University of Michigan Library, 2009), http://hdl.handle.net/2027/spo .did2222.0001.306 (accessed 8 December 2012). Originally published as "Laboureur," *Encyclopédie ou Dictionnaire raisonné des sciences, des arts et des métiers* (Paris, 1765), 9:148.

81. Antoine Laurent and Marie Anne Lavoisier, "Observations sur les familles d'Indiens amenées en France par M. de Suffren," in *Les Oeuvres de Lavoisier*, http:// www.Lavoisier.cnrs.fr/ice/ice_page_detail.php?lang=fr&type=text&bdd=koyre_Lavoisier &table=Lavoisier&bookId=314&search=no&typeofbookDes=Memoires&facsimile=off& cfzoom=1.5&pageOrder=1 (accessed 1 December 2012).

82. Steven Shapin, "The Invisible Technician," *American Scientist* 77 (1989): 554–63.

83. Antoine Lavoisier to municipal officers of Blois, 16 March 1789, in *Oeuvres de Lavoisier*, 6:34.

84. Marisa Linton, *Choosing Terror: Virtue, Friendship, and Authenticity in the French Revolution* (Oxford: Oxford University Press, 2013), 52–53. Some nobles ignored this prohibition, but electors of the Third Estate were much more inclined to abide by it.

85. Ibid., 58.

86. "Copie de la déliberation de l'Assemblée de Messieurs du Clwb de Blois," sent by Petit de Villanteuil to Mme Lavoisier, 25 March 1789, in *Oeuvres de Lavoisier*, 6:35.

87. Antoine Lavoisier to municipal officers of Blois, 19 April 1789, in *Oeuvres de Lavoisier*, 6:38.

88. Antoine Lavoisier to municipal officers of Vendôme, 15 August 1789, in *Oeuvres de Lavoisier*, 6:62.

89. Antoine Lavoisier to the Arsenal Section, 29 June 1790, in *Oeuvres de Lavoisier*, 6:145.

## CONCLUSION

1. J. A. N. Condorcet, *Oeuvres de Condorcet*, ed. Arthur Condorcet-O'Connor (Paris: Firmin Didot Frères, 1847–1849), 2:637–38.

2. Claire Cage, "'Celibacy Is a Social Crime': The Politics of Clerical Marriage, 1794–1799," *French Historical Studies* 36, no. 4 (2013): 601–28.

3. Paul White, "Darwin's House of Science" (paper presented at the International Conference for the History of Science, Technology, and Medicine, Manchester, UK, 2013).

4. Sarah Horowitz, "The Bonds of Concord and the Guardians of Trust: Women, Emotion, and Political Life, 1815–1848," *French Historical Studies* 35, no. 3 (Summer 2012): 577–603.

# INDEX

Page numbers followed by the letter *f* indicate figures.

Du Châtelet-Lomont, Florent-Louis, 109–15, 197n30
Duhamel du Monceau, Henri-Louis, 139–41
Du Marsais, César Chesneau, 17–18, 20–21, 108
Duperré, Marie-Soulange, 33
Du Pierry, Louise-Elisabeth-Félicité, 43, 45, 65
Dupin de Francueil, Suzanne, 13
Du Ponceau, Doigni, 25
Du Pont de Nemours, Samuel Pierre: eulogy for himself of, 154–55; experimental farm of, 143, 145; relationship with Marie-Anne Lavoisier of, 15, 38, 151

*École d'Agriculture* (Duhamel du Monceau), 139–41
education, 3, 32–33, 102–35, 195n6; of collaborative wives, 46; of Condorcet's daughter, 102, 131–34, 201n109, 201n113; in convents, 115, 119–20; of d'Epinay's children, 126–31; of Diderot's children, 32–33, 116–25; of Du Châtelet's children, 109–16; French Revolutionary reforms of, 105, 131–32; of girls, 107, 109, 115–34, 201n109, 201n113; Jesuit forms of, 105–6, 112; malleability of children and, 103–4, 127, 131; on patriotism and civic virtue, 103, 106–12, 134; pedagogical texts on, 109, 111, 113–15, 126–31, 135, 197n29, 200n94, 200n97; Rousseau's influence on, 106–8, 127; study of anatomy in, 120–21, 198n74; utilitarian goals of, 108–12, 116, 134; women's roles in, 102, 108–17, 126–31, 135
"Education" (Du Marsais), 108
*Emile, ou de l'éducation* (Rousseau), 13, 182n90; pedagogical program of, 106–8; on smallpox inoculation, 84; Sophie's passivity in, 134; on women's domestic roles, 62–63
*Encyclopédie* (Diderot and D'Alembert), 6; on education, 108; on farm laborers and peasants, 161; on smallpox inoculation, 77, 94; on sociability and public utility, 17–18, 20–21; on women's domestic obligations, 62, 188n83
English agriculture, 139–40, 143–44
enlightened love, 82–83, 90–93, 96–97
Enlightenment thought: on agriculture, 138–41; anticlericalism of, 105–6, 119–20; on children's education and upbring-

ing, 103–10; commitment to public debate and sociability in, 7, 16–18; on the *honnête femme*, 124; inoculation debates in, 76–79; marriage debates in, 14–15, 166–67; on military reforms, 113; pairing of reason and sentiment in, 6–8, 81–84, 90–93, 96–97, 129–30, 147–50, 159–60, 166–68; physiocratic economics in, 145–47, 203n42; popularity of nature in, 141; self-identified participants in, 10–11; on sentimental domesticity, 19–20, 166–68; social reform goals of, 3–6, 12, 53–54, 168; on women's intellectual work, 40–41, 61–63, 183n2, 188n83; on women's roles in society, 122–25, 131–35, 199n79, 201n113
*Éphémérides* (Lalande), 43, 65
"Essai sur la société des gens de lettres et des grands" (d'Alembert), 24
*Essay Concerning Human Understanding* (Locke), 103–4
*Essay on Phlogiston and the Constitution of Acids* (Kirwan), 49
Estates General, 163–64, 205n84
*Expérience sur la respiration de l'homme au repos* (Marie-Anne Lavoisier), 50f
*Expérience sur la respiration de l'homme exécutant un travail* (Marie-Anne Lavoisier), 51f
experimental farms: agricultural reforms of, 138–51, 203n42; enlightened benevolent landlords of, 151–65; manual labor at, 161–62
*Experiment on a Bird in the Air Pump, An* (Wright), 120, 121f
extramarital relationships. *See* unmarried households

*Fable of the Bees* (Mandeville), 115–16
family life, 166–68; cultivation of sociability in, 18–21; Fontenelle and Perrault's ambivalence on, 21–24; gendered division of labor in, 40–43, 62–63, 107, 109, 188n83; language of sentimental domesticity on, 19–20, 166–68; in metaphors of social reform, 1–2, 4–6, 137; patriarchal focus of, 41, 64–65, 68–69, 80–81, 137; production of knowledge in, 9–10, 12, 14, 40, 43–60; Rousseau's views on, 13, 62–63, 182n90; sentimental love in, 5–6, 9, 19–20, 40–42, 61–69, 71–72, 81–84; smallpox inoculations and, 3, 70–101. *See also* col-